Elizabeth Tuckett

Beaten tracks

or, Pen and pencil sketches in Italy

Elizabeth Tuckett

Beaten tracks
or, Pen and pencil sketches in Italy

ISBN/EAN: 9783337013967

Printed in Europe, USA, Canada, Australia, Japan

Cover: Foto ©Thomas Meinert / pixelio.de

More available books at **www.hansebooks.com**

BEATEN TRACKS

OR

PEN AND PENCIL SKETCHES IN ITALY.

BY THE

AUTHORESS OF 'A VOYAGE EN ZIGZAG.'

'*Of making many books there is no end.*'

LONDON:
LONGMANS, GREEN, AND CO.
1866.

PREFACE.

THESE letters were written to a sister in England, during a journey in Italy early in the present year; the narrative is so slight a one, that it would hardly be worth presenting to the public, if it were regarded as anything but a description of the sketches, which, in their turn, serve as illustrations to the journalisings, the drawings having been often made on the margin of the original letters.

As a piece of patchwork, the book may serve to recall to old travellers, or picture to new ones, the quaint mixture of romance and bathos that belongs to modern wanderings. The kind reception given to our 'Zigzags,' makes us hope that this, though a more rectangular journey, may be welcomed at Christmas firesides, and that some of the superabundant charity which is supposed to pervade such an atmosphere may cover the deficiencies of our book.

November 1865.

LIST OF ILLUSTRATIONS.

Florence . . . . *Frontispiece*	
	TO FACE PAGE
Jardin des Tuilleries—Parisian *Bergères*—Paysanne—La Mode (April)	5
Quai Dessaix—The Rhone—Tarascon	7
Fau-Choo—At the station, Toulon—Fréjus—An Exile .	15
Marseilles—An Ayah at the Hotel, Marseilles—Toulon .	19
Toulon	21
Market at Cannes	25
Cannes—Cannes, 6.30 A.M.	27
A little Breeze on the Promenade Anglais, Nice—Spanish Pedlars	30
Mentone	35
Washerwomen, Mentone	37
A Bough from an Orchard—Birra !—On the Riviera .	39
Baby-Show at Alassio	45
Our Dinner—Sketches on the Corniche—Pescatori—Charcoal Carriers	47
Savona—On the Riviera—Alassio . . .	49
A Genoese 'Peeler'—Genoa	51
Genoa	61
A Sketch at Porto Venere—On the Riviera . .	65
Spezzia—By the Roadside . .	67

TO FACE PAGE

Farina Francesco—Lord Byron's Boatman—Porto Venere—
   Garibaldi's Prison 69

Porto Venere 71

Carrara—In the Quarries—Carrying the Marble—Carrara,
   Pistoia—Pistoia Chestnuts 75

A Carrara Atelier—Sarzana—Pisa, Spezzia: Italian Chambermaids—Pistoia Tower—Pisa 81

Bagni di Lucca—The old Custode—Ponte 89

In the Bargello—Cascine; the Piazzone 105

Fiesole—The Dante Festival—May-day—The Ball at the
   Uffizi 121

Palazzo Vecchio—Dante Festival—Monte Oliveto . . . 141

At the Gate of the Cemetery—Hiram Powers—Ascension Day 163

Grillo—Ascension Day—In the Lanes 175

In the Piazzone—Ponte alla Caraia 185

Fiesole—Above Fiesole—Castello Pucci 193

Santa Margarita—The Valley of the Emma, from Santa
   Margarita—Lung' Arno. 201

Elise—Florentines 217

Carolina Bernardi—Waiting for the Bride 223

Bologna—Monsieur Polichinelle 237

Bologna—Lake Como 241

Como—A Travelling 'Troupe'—Young Russia . . . 247

Garden Serbellone—Bellaggio 249

Bellaggio 253

Airolo 259

Crossing the St. Gotthard—The Devil's Bridge—At the
   'Goldener Schlüssel' 261

Lucerne—'Ockheimer, Garçon!'—Basle, our Omelette—
   Posting the Last Letter. 271

BEATEN TRACKS.

I.

Paris, Hôtel du Louvre: April 4, 1865.

As usual, the journey has been a very easy affair, so far; or rather, I ought in honesty to say, as far as Calais. We had the customary dinner at the Charing Cross station, took leave of F., who travelled with us to town, and had a long gossip with P., who came, as last year, to wish us *bon voyage*.

There is always something to me especially comfortable in the sense of being settled into one of those cosy 'South Eastern' carriages, and in having fairly made the start. All the bustle and worry of leaving is over; our spirits are toned down into a lazy content that generally preludes a luxurious doze. We seem to run easily along the rails till the sound changes to a hollow crash with a quiver in it, and we are passing a bridge with the lights of the great city over us, and beside us, and beneath us, in indescribable confusion; and we glide like spirits between and over rows of houses; then, with a little roar, pass into darkness and out again into a misty glimmer of sky and river, and tall chimneys, till we hardly realise which are phantoms and what is reality, or whether it is not all a dream, and we are beginning to feel hazily uncomfortable about our luggage, which will haunt the feminine mind even in repose; remembering that a porter told us none of the boxes were locked, and that the salt water had got to our bonnets; for somehow we seem already to have made the crossing, only we have landed at the wrong place, and cannot find any train going to Paris, when suddenly a voice says, 'I think he's in the right, you know, and Lord Palmerston is bound to support him'— and we are once more in our corner of the carriage, rushing steadily through the darkness, and a lamp is burning, and my father reading a last instalment of the beloved 'Times,' in the pauses between fragmentary discourse with two gentlemen opposite. We let down the window and put a head out into the night: there is very little wind, and somebody says, 'We shall have a first-rate passage;' and then comes a glimmer of water and a sense of space and sea, and lights again, and dark blocks of building growing out of breadths of fields and downs that have been running past us in long stretches of shadow through the night.

There is a great red ball like an angry eye, and a black hull and masts, and a tall funnel down below us, and we come to a stop suddenly on the long pier—all the carriage-doors are flung open, and porters hurry about shouting to each other; a soft gusty breath from the water flutters our dresses and makes us fold our warm wraps more closely round us, as we descend the wet slippery steps and the plank across the water, making our way among coils of ropes and sacks and various piles of luggage, to the little lantern on the deck through which we pass into the depths below.

How well one remembers former crossings, and the indescribable smell of salt water and brandy and broiled bacon, that always seems to come with a whiff up the brass-nailed stairs, down which one slips, thankful to reach the firmer ground of the cabin.

The night was glorious, still and clear, and warm enough, the steamer not overcrowded; a few French women kept up their pretty musical tinkle of talk between spasms of fearful sea-sickness, which was all an affair of imagination, as the water was smooth as a millpond. There was a white cockatoo in a cage on the table, who swung about miserably, and said 'Pretty Poll' in the most deprecatory of voices. C. and I lay down on some upper shelves, covered up heads and ears, and rested capitally, though we did not get much sleep.

On landing at Calais, we found ourselves at low water, ever so far from the station; but we were bundled into an omnibus, parrot and all, and rattled off to the *embarcadère*.

The train was, as usual, crowded, and it was hopeless to think of securing any extra space.

We made a good supper in the waiting-room, C. and I enjoying some smoking hot bread and milk and cold chicken, and watching with much amusement the unfortunate victims of *maladie de mer*, who regarded our choice with faces of horror and astonishment over their weak soup or *petits verres*. There was a large fire blazing in the grate, which looked cheery and Englishlike, and we made ourselves very warm and comfortable before starting again with three fellow passengers, an old English officer, a very small boy, and a hideous little Frenchman, who looked so ghastly that C. immediately decided he was infectious, and refused to enter the carriage; but our fears calming down, we all subsided into our places, and got through the rest of the night somehow. The discomforts of the journey always begin for me at Calais. I can never grow reconciled to a railway carriage the wrong side of midnight; but the hours passed, and we managed to sleep a little.

We were most comfortably warm, and the hot-water tins were changed once during the night. After the severe cold we had experienced so short a time ago, this milder weather is a great boon.

The little boy and I fraternised over some French plums; he ate an apple about 5 A.M., and a bun which he had sat upon during the first part of the night, and seemed much comforted by the refreshment they afforded. He was very good and quiet—poor little fellow!—and the old gentleman was very civil. When the little Frenchman went to sleep, my father shut up his window, and the moment *he* began to doze the Frenchman always woke up and opened it, my half-waking consciousness realising our companions as so many puppets with invisible strings. So the night wore on, one's eyes lazily unclosing now and then at the click of the window, as it went up or down, and as the day dawned, catching a glimpse of the poplars through the frost on the glass; pale gleams of colour, then a great scarlet ball, and it was broad day—a bright, fresh, exquisite morning—and we were in Paris; the luggage was passed, and we were rattling through the streets in a little omnibus to this pleasant, home-like hotel.

C. and I, who have never been in Paris so early in the year before, were much interested in the winter costumes of the people—the women all wearing scarlet, or blue, or purple hoods (*capotes*, hood and cape in one), pretty and very becoming, instead of the white caps or nets; many of them with fur collars over their print dresses, instead of cloak or shawl.

We found this hotel far from crowded, and secured good rooms close to the great staircase, and facing the Louvre.

We have an enormous room, properly a salon, with two beds in it, and my father's room adjoins it. We washed and dressed, and then breakfasted by a blazing

Jardin des Tuilleries.

Paysanne.

la mode.
(April.)

Parisian Bergères.

wood fire in the salle, numbers of English and Americans in the rooms, mostly coming up from Nice or Italy; and all rather shabby and seedy-looking, as though they had come to the end of their clothes!

While my father was enjoying the papers we read and slept, going out together about twelve. I cannot describe to you how lovely the whole place looked; the weather has been very cold, and this sudden burst of spring sent everybody swarming into the open air. The Tuileries gardens were crowded with nurses and beautiful fat babies and little dots of children in white, with woollen hoods and white boots, tumbling about under the leafless old trees, like spring snowdrops and crocuses, the real flowers of the place; and older boys and girls in cloth and velvet, bright blue and violet with rich fur edging, and coloured and tasseled Hessian boots to match; artisans in blouses sitting about with very small sons and daughters, like little sacks tied up at one end, and ornamented with ribbons on the tiny caps, all so happy and chirping like grasshoppers; girls on their way to and from work, feeding the tame sparrows and pet pigeons with crumbs saved from their breakfast; fountains plashing in the sunshine; the bluest of skies, with a soft purple haze, far away through which the Arc de Triomphe glimmered like a white star.

We drove round the Boulevards to see the new buildings, and what had been finished since last autumn, through the Parc de Monceaux, passing more nurses and babies and little family groups, by the Greek church, and so through the Avenue de l'Impératrice to the Bois—where we drove for hours, basking in the sunshine and rejoicing in the air after the long wearying night in the train. Thousands of people were out on foot or in carriages, boating parties rowing on the lake, whole families camping out in the woods or hunting for violets, and ladies and gentlemen lounging on seats under the trees.

About four o'clock the crowd of carriages rapidly increased, and the show was really magnificent. I never saw anything like it in Paris before. Madame Rothschild, like Miss Kielmansegge, all real gold, made a most startling display in a carriage, like a Lord Mayor's, all over bullion! Some lovely women passed, but the Empress did not appear. She is not well, and seldom drives, they tell us.

Wednesday, April 5.

Whilst we are spending an hour or two indoors I shall add a little to my letter. It is so wonderfully warm and summerlike, that we are thankful for an open window in our great room. We all slept soundly last night, and are well rested. After breakfast we watched the guard being changed and paraded before the palace, to the great amusement of the little prince who superintended matters from a window.

We spent some time again in the Tuileries gardens, feeding the birds and talking to an old French peasant woman evidently fresh from the country, who accosted us with many enquiries as to the meaning of the wonders around her, innocently unconscious that she was receiving her information at second-hand from foreigners.

We drove to the Quai Dessaix and sketched some of the old women busy over their delicious flowers, and on to the beautiful Gothic church of Ste. Clotilde which looked lovelier than ever—such a glorious rich light and colour on the pillars and bas reliefs; and went again to the Bois, amusing ourselves with the pretty faces and costumes. Madame Rothschild and the Baron as the 'golden dustman' passed again, and the same ladies and carriages, and babies and little pet dogs. We drove to the cascade through the Pré Catalan, where there is a *laiterie*, round which dozens of voitures were drawn up, while the great ladies and their children sat on benches in a yard, and drank fresh milk served to

Quai Dassaix.

The Rhone.

them by Parisian bergères in Swiss hats and short petticoats—a ludicrously artificial costume—making quite a little fête out of the performance. It was delightfully Arcadian, like poor Marie Antoinette's dairy parties, or a highly finished picture by Watteau.

How impossible it would be for London ladies and gentlemen to drink milk gracefully on the banks of the Serpentine! But the French have implicit faith in themselves, and their *mise en scène* is quite perfect.

The coachmen look wondrously comfortable in their high rolled collars and tippets of magnificent furs; but spring fashions are rapidly blossoming out in the sunshine. The smallest bonnet I have seen consisted of a bunch of wheat or oats, a slender sheaf, laid on the top of the head and tied on with a broad ribbon. On the box of one very grand carriage the footman was holding three large wooden hoops for the children, who were inside; and the numbers of nurses and infants driving out were quite wonderful.

The people were amusing themselves as usual in the glades of the wood, and we came suddenly on a group under the trees (a grown-up one), playing *puss in the corner*, which seemed to necessitate a good deal of kissing of one very pretty young woman in a red hood.

Our drive had a most disastrous ending. We dismissed the carriage, which was an ordinary hired one, at the corner of the Palais Royal, and five minutes afterwards discovered it had driven off with our invaluable waterproof cloaks, and my sketch-book, deposited safely in its head. As we had not kept the man's number, there was no hope but in his honesty, and as he knew that we were at the Louvre hotel, he may have the Christian charity to return our precious possessions. Alas! it is a very serious trouble, an unlucky beginning; and what if we go south only to encounter a steady downpour, incessant rain, to which we may be exposed without our faithful protectors!

I know how deeply you will sympathise in our distress. My father has not uttered one word of reproach, or even 'pointed a moral' in reference to the future, and promises to give us any amount of wraps to be bought in Paris; but our beloved *tweed* is an unknown luxury here, and we are retiring to rest with heavy hearts.

We have made these two days in Paris, as you will see, as resting as possible, before the long journey to Lyons. We have some new '*Tauchnitz*' and a pâté de foie gras, for the eleven hours in the train. There is nothing very new in the shops; swallows still a favourite decoration, and dogs' heads also. I bought a wooden fan to paint a bird on; and we have a goodly store of bonbons.

My father walked through the market this morning, and brought us two red rosebuds, our first roses this year. They have rebuilt the old fruit market, and he had quite a hunt to find our little woman, who gave him a rapturous greeting, and pressed him to buy some of her pears or oranges; and on his return through the Rue de Rivoli, her husband passing in his cart, recognised ' monsieur,' sprang down, and seized upon him, explaining 'with effusion' the whereabouts of their new place of abode.

I don't like to think we must wait till we reach Nice for your letters. . . I have written in great haste, so forgive all mistakes and repetitions.

Paris, April 6.

We are just off for Lyons; *no cloaks*.

II.

Grand Hôtel de Lyons, Lyons: April 7.

WE have still the same glorious weather; cloudless blue sky and a pleasant cool air this morning, which is very refreshing after the intense heat yesterday, increasing rather than lessening towards night, and making us realise in the crowded railway carriage that summer had begun all at once.

We were early at the station, and were able to secure comfortable places, which it is not always easy to do in these express trains, where there is no second-class to take off the pressure of travellers. We were glad to find that our luggage, which had been pronounced at London Bridge very much over the authorised weight, and charged for, of course, accordingly, was not likely to prove so expensive a luxury in future, though whether this arose from any mistake in England, or from a more generous allowance as to quantity on the French lines, we could not ascertain. Our carriage gradually filled up; and it was amusing, in spite of one's pity for them, to watch the anxious late arrivals hurrying hither and thither seeking for places: amongst them was an odd-looking invalid, with his head bound up in a towel, dragged about by a tall ghastly companion, who at last settled him into another compartment after making some vain attempts upon ours, and from this at the last moment a man and maid-servant laden with cloaks came rushing, and filled our last two seats, explaining breathlessly to us that the caretaker of the sick man had announced that his friend was suffering from scarlatina—a fact that effectually emptied their compartment. Of course people may sometimes be compelled to travel under such circumstances, but they should at least avoid endangering others, and take the coupés which are so convenient for invalids, though rather more expensive than the ordinary places. It is no good to be too nervous about possible infection, but at the same time it would not be pleasant to discover, with no station for an hour or more, that your next neighbour had the small pox.

Our other fellow-travellers were an English gentleman, who was silent and quiet most of the way; but became very pleasant and communicative towards evening; a Frenchman, who looked like a courier; and a tall and very gentlemanlike young priest, with whom, later in the day, we had much talk. E. and I amused ourselves with our 'tatting' for some time, whilst the maid stitched industriously at a roll of worsted work; and then we varied the time with our books, and looking out at the very pretty country we were passing through, though, the vines being only represented by dry sticks, one lost all that will in two or three months make the plains and the hillsides so green and lovely. Many of the little towns were very quaint and pretty, with fine old churches or ruined towers and châteaux on the hills. The chief drawback was the dust, which quite rattled against the carriage, and covered us all over with shiny specks; but it all shook off very easily, and we refreshed ourselves occasionally with a small eau de Cologne bath, washing face and hands with it, until our English fellow-traveller warned us solemnly of its injurious effect upon the skin; but I cannot say we have seen any serious results as yet, and the comfort at the time is very great.

At Tonnerre we had seven minutes to get out and shake ourselves, and then two or three hours more to Dijon, where at six o'clock we found a table d'hôte dinner ready, and twenty-seven minutes to eat it in, so that we made a good meal, and started again well rested and refreshed, and able thoroughly to enjoy the remainder of our journey, with the lovely evening lights and shadows, an orange and purple sunset, and a sort of suspicion of the Jura far away; and, as it got too dark to read or work, the priest put away his book, and the Englishman began to thaw, and we were soon in the midst of most amusing and pleasant talk about England and France, the French gentleman speaking English rather shyly at first, but as though he enjoyed it, and expressing most warm admiration for our country and an earnest desire to visit it. The feeling of dislike to England is, he told us, still very general, especially in the more remote parts of France, where the *entente cordiale* has hardly yet penetrated, and the old jealousy of our wealth and prosperity still exists.

As a French gentleman in the Pyrenees once said to my father: 'If we meet Englishmen we like them, but as a nation we are not fond of you. You are richer than we are; you are more successful; we are a more inventive people, and you carry out our inventions; and we have never forgiven Waterloo!' It is curious how comparatively little foreigners seem to know of English life and ways. Our friend the priest asked many questions that showed how strange it would all be to him if he were ever to cross the Channel, and he was surprised and pleased to find that so many of the French books of the day were read and appreciated in England, and that we could talk about them with him; but he much lamented the inferiority of the tone of much of their literature: 'We have so few books that we can give to our sisters.' He exclaimed with such pleasure when 'Eugénie de Guérin' was mentioned, saying that M. de Lamartine declares, 'Of all the books of this nineteenth century I love that the best.'

We told him how much we had been interested by Mr. Musgrave's book 'Ten Days in a French Parsonage,' and his history of the Curé's visit to him at his English rectory, and their respective first impressions of French and

English parish work and life. He seemed astonished to hear that our clergymen dress so much like other people, and asked whether he might wear his cassock if he came to England; he should not mind being laughed at by the small boys if the police left him unmolested! He is related to several English families, and was a most thorough gentleman, and so simple and amusing, saying, 'I like to laugh sometimes;' and when we all separated at Lyons, 'I have had such a very pleasant English lesson.' He was evidently utterly ignorant of our most common customs; and it was strange to see this in an educated gentleman evidently belonging to the upper classes of French society, making one realise how narrow were the education and the habits of thought impressed upon the young men in Roman Catholic seminaries.

We had a long drive in an omnibus from the station to this very good and thoroughly modern hotel, where we were glad to find rooms on the first floor, and to go to bed about twelve o'clock, after having some tea in the salle. Lyons is so different from the dismal town I fancied it, for we drove last night through very fine streets and squares, and it all looked so bright and open; but no doubt we have seen the best parts of the city.

Marseilles: Saturday, April 8.

We are comfortably established in the new Hôtel du Louvre, and before going to bed I must add a little to my journal. Yesterday morning, after I wrote, we walked through the principal streets of Lyons, visited the handsome Bourse, where the business for the day was just over, and also a church, in which I think there was nothing of especial interest. The shops struck us as rather poor, but some wonderful ribbons tempted us into one of the handsomest of them. These specimens of their manufactures were really beautiful, as well as curious—figures, coats of arms, and every imaginable device being introduced into the silk in all varieties of colour and form.

At two o'clock we set off in an open carriage with good horses for a long drive, first up the steep ascent to the fortress commanding the town—a very unpopular check upon any expression of rebellious feeling in Lyons, which could be swept from end to end by its great guns. We met a number of soldiers, hot and dusty already, leaving this garrison, I suppose, for some other. After winding up between the massive walls of the fortifications, you reach the little town which they enclose, with its strangely narrow streets and small shops, and see, occupying the most elevated and conspicuous point above the city, the church of Notre Dame de Fourvières, crowned with an immense gilded statue of the Virgin, which is supposed to protect the city, and is an object of extreme veneration to the Lyonnese. The interior is rich in votive offerings, but the building itself is plain and uninteresting. Close to it we encountered numbers of children, evidently belonging to charity schools, and come to celebrate some fête in this favourite church; they were all distinguished by a badge, medals, coloured ribbons or dresses, and looked neat and very happy. Many of them were in blue, the Virgin's colour, and probably belonged to an establishment attached to this church.

From this height we had a beautiful view over Lyons and the Rhone and Saone, and far away the Alps and Mont Blanc were clearly to be distinguished in spite of the light clouds hanging about them. The day was perfect and summer-like, with an exquisite blue sky and soft balmy air.

We drove down again by a still steeper road than that by which we had ascended, and saw the city itself very thoroughly, passing many pretty fountains and open *places* where nurses and children were congregating.

Then our coachman took us to the Bois de Boulogne, one good result of the great floods, as the ground outside the town, which was so desolated then, has been made into a really charming park, with drives and walks and a lake; one portion is formed into a zoological garden, and it is all open to the public, and well planted with quantities of cedars and ornamental trees, amongst them unusually fine specimens of *Wellingtonia gigantea* in great numbers, but almost all sadly browned by the smoke. There were not many carriages, although we were told it was the usual time for driving there; but possibly the heat made people choose a later hour.

This morning we were up at half-past five, waking to another glorious day; and after breakfasting at half-past six we drove to the station and joined the express from Paris. We enjoyed the journey extremely, as the line passes through such lovely scenery, the chief feature in which was the Rhone, with its numberless little towns and castles and bridges on one side, and far away on the other were the Dauphiné Alps and Mont Blanc, giving us terrible *maladie du pays* from the longing to be amongst them again. Then, too, we had the bright pink and red almond and peach blossom, and the first olive trees and ilexes, and little villages with their flat Italian-looking roofs and colouring; and at Avignon and Arles a glimpse of the old Roman remains; and so at last we reached the strange, barren, desolate-looking, marshy land about the mouths of the Rhone, having turned away from the river itself when we left the more fertile country behind us. These *camargues* are very curious, so stony and arid that there is hardly grass enough to support a few sheep, and the one small village in the midst has the most forlorn and dismal effect, and one feels that each passing train must be a welcome sign of life and civilisation. It is truly a desert, and we even saw a mirage at the foot of the hills, repre-

At the station, Toulon.

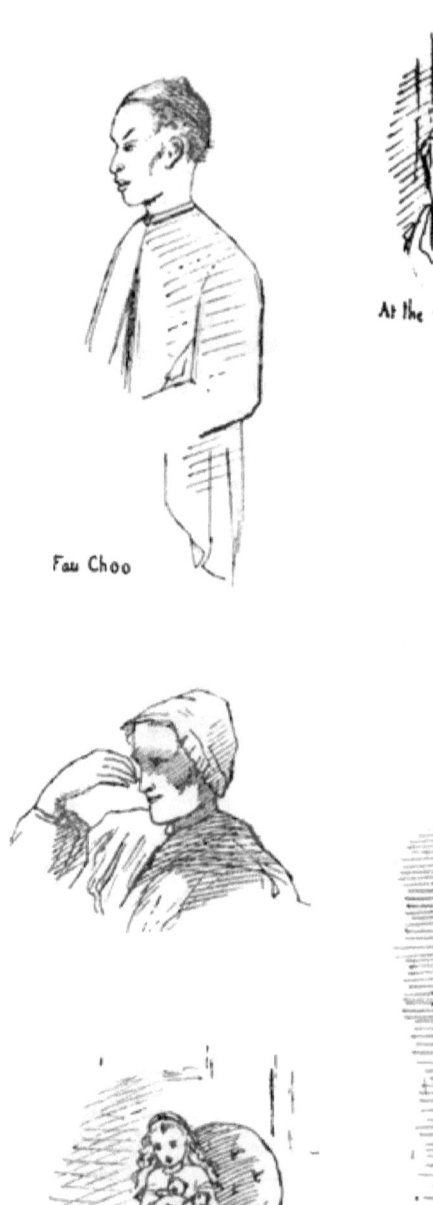

Fau Choo

Frejus.

An Exile.

senting an imaginary lake, whilst nearer us lay a small one of veritable water of the most extraordinarily brilliant green.

Suddenly the first view of the Mediterranean opened upon us, blue and lovely, and almost realising one's anticipations, and then soon we were passing into Marseilles, through its suburb of tiny villas and gardens, which form such a marked and peculiar feature in the immediate environs of the city, none but the very poor feeling content without a small country residence, although often these *bastides*, as they are called, seem part of the town itself; and the result is most strange in the endless varieties of minute houses, looking almost as if designed for dolls, and exhibiting every sort of architectural freak and fancy on the part of their respective owners.

This is a very fine hotel, with a large central court, round which are the salle à manger, salon, and the magnificently decorated room where breakfasts are served, and which is also used as a café or restaurant. It is amusing to watch the people coming and going, travellers from all parts of the world, including little children from India, who are rejoicing over some small fat puppies which are inclined to be sociable, and run in and out of the bureau all day long.

April 9.

This morning we walked to the little English church, or rather room, in a quiet street, which we found warm and very crowded. A conscientious clergyman did his duty admirably, though the sermon was rather a failure— a dry analysis of the plagues of Egypt. He was to preach again at three, and again in the evening on board one of the Peninsular and Oriental steamers; and it must require a very great mind indeed to be original under such constant pressure.

On leaving the church we consulted an English gentleman, who was turning in the same direction as ourselves, about the road to the chapel of *Notre Dame de la Garde*, on which he joined company and we climbed up together, my father enjoying a talk with him over places and scenery, and comparing notes on hotels &c., as he had just left Cannes and Nice. The weather was perfect, clear and bright and cool, so that we saw everything to perfection. A new church has been built on the site of the old one—a rather imposing structure from below, but an ugly modern affair in reality, half Italian in style, with stripes and blocks of coloured marble. All the little old votive offerings are gone, except the pictures, which seemed to us very touching, notwithstanding their oddity and ludicrous violation of all rules of art.

A shipwrecked boat with the Madonna coming out of a cloud to rescue sailors, who turn to Notre Dame de la Garde as their especial patroness; a little sick child on a bed, with a weeping father and mother in agonised supplication, and the heavenly mother shining down on the little one's face; a man falling from his horse, an overturned omnibus, a broken leg, and so on through the whole round of human misfortunes—the truth of the belief in each giving it a sort of humble beauty in spite of its grotesqueness.

We sat for half an hour in the porch gazing at the wide expanse of blue sea, the islands and the Château d'If gleaming softly out of the water, and the cliffs and houses white and glistening in the sun. Many people passed up and down while we waited for the sacristan, who was at dinner and must not be disturbed. I don't know whether any especial blessing is entailed by a visit to the church and the climb, which seemed rather to try the lungs of several portly citizens, who puffed and panted up the steps behind us. We sat on an unfinished wall watching each fresh head as it appeared below us—now a group of blue-capped chasseurs, then a woman with a gay kerchief bound over her hair, her baby in her arms, and her four-year-old child struggling up behind her with a wonderful effort and stretch of the fat legs, as the small thing achieved each fresh mountain of a step, its little body nearly toppling over with the tremendous exertion, while its dignity declined to go on all-fours ; a few sailors, sending a cloud of smoke before them, warning us of their approach ; two or three young girls, fresh and trim in their Sunday finery ; and groups of the better class, shopkeepers or merchants ; and dried-up old women, like nuts, shrivelled and warranted to last for an indefinite period.

We walked down from the church and through streets so steep they seemed to have been intended for staircases, only the steps had been forgotten, and so back to our hotel, and luncheon, and a quiet rest in our rooms ; starting later in the afternoon for a little drive before dinner.

We are charmed by Marseilles, which, from everybody's account, we had fancied must be a dusty furnace, heated seven times for the reception of those unhappy travellers who might happen to be detained there by wind or weather. We have seen it at its brightest and best, and have, no doubt, been so far very fortunate ; *no* dust, and no great heat, for there has been a fresh breeze all day. The streets were thronged with people quietly walking about, bright and happy-looking, and handsome for the most part; they swarmed all over the broad streets in true southern fashion, and sat on doorsteps, and formed little colonies of chairs in front of the shops. The awnings struck us as something new, like flapping valances of cotton, hung round wires, not strained on or over them, and very picturesque in colour. The 'Bois' through which we drove had the oddest effect. Rows of closely-cropped plane trees, looking like Dante's tree-men, in the agonised twist of such stunted branches as had escaped the pruning-knife. It was difficult to believe they could shoot out into branches and leaves in time for summer, or afford any shade for years to come to the sun-stricken citizens.

I do not know how old the trees may have been, or whether the whole avenue is a recent invention, like those in the Palais-Royal or the Boulevards in Paris. Do you remember the scene with the irate Frenchman in one of *Cham's* old books of ' Croquis ? '

Le Gardien—' Vous réclamez cet arbre ? '

Le Bourgeois—' Ça un arbre ! c'est-à-dire que je reconnais ma canne que j'ai perdue il y a huit jours, et que je retrouve maintenant plantée dans le jardin du Palais-Royal. Rendez-moi ma canne, je veux ma canne.'

The carriages and people were very amusing, and the scene most gay. We drove for some time along the *Prado*; such a glorious sea road, with the blue water coming almost to our feet, and high cliffs on one hand crowned with every sort of queer little house; with people drinking coffee in balconies, and women and babies in summer-houses on the walls, coloured pink and yellow and vermilion like the dwellings above, gay with their green jalousies; and so on by the harbour and shipping, where the yards of the lateen sails made pleasant angles amongst the northern rigging, and the great yuccas and cypresses and palms, peeping between the houses, made us realise we were coming south.

But I should only weary you if I told you half the sights which amused us as we passed on; and Mr. Ormsby's charming ' Autumn Rambles ' contains so brilliant a description of the quays and shops, and sights and sounds of this strange modern Babel, where every nation and language has its representative, that I can only feel that everything possible has been written about it, in the cleverest possible way, and that as you know so well his picture of it all, you must not be troubled with mine.

We reached our hotel in good time for the six o'clock dinner. My opposite neighbours were some very genial

well-bred Americans, just returning from Spain, and *en route* for Rome. We had much pleasant talk; and hope to meet again, possibly at Venice. On leaving the table, we were joined by the C.'s in the salon, and had an amusing gossip.

This hotel is well built; our rooms are very large and most comfortable. There is an immense hall like a well in the centre of the house, glazed over, with chairs and plants and flowers; and the salon and breakfast-room, which serves also as a restaurant for the townspeople, are really magnificent.

I must not forget to tell you that we passed the gates of the Imperial palace, a grand building overlooking the sea, and the new Imperial street, which cuts through a poor dismal quarter of the town, and turns some most miserable cottages into great houses, like a transformation scene.

They have a Chinese waiter here, and a little mite of a dog, with two lovely pups; soft round balls of love and good temper, with no particular heads or legs at present. They are generally to be found very happy and content in the arms of one of the children, who are everywhere in the hotel. Many pale-faced little Anglo-Indians, with solemn ayahs, with an exile's yearning and a Hindoo's fatalism and patience in their soft sad eyes. I came upon one suddenly a few moments ago, the strangest-looking creature, like a broom wrapped and swathed in old yellow and red silk. My father, of course, fraternises with all the children everywhere, and he and the little girl and the dogs have become quite intimate, though he startled her with awful suggestions of possible danger to her pets, in reference to puppy-pie and Fow Choo, the waiter! This is a patchwork letter, nominally from C., with my Marseilles history as a bit sewed on, but you will not care, as it is all good news.

We start soon after eight to-morrow for Toulon, stop

there for two hours to see the *galériens*, if possible, *à propos de Jean Valjean*, though I fear it will be a dreadful, pitiable sight, then on to Cannes for one night, and the next day to Nice.

III.

Cannes, Hôtel Gonnet: Monday, April 10.

WE are sitting in my father's room, looking out on such a scene of loveliness, the calm long quiet waves lapping the shore, the town on one side gay with little twinkling lights, and on the other the curve of land in soft shadow, with a broken ring of silver where the edge of the waves springs up into the moonlight.

No words can describe the beauty of our journey today, as the evening drew on and a bright rosy light lit up the Estrelles mountains—the sea one soft turquoise, the sky almost too clear for colour, the distant hills fading away into the tenderest pink tintings, the farthest tipped with snow, and every line and ridge cutting clear against the light. The valleys and plain near us were dotted with olive trees, vines, and the great southern pine trees, with here and there a stray palm and yuccas and oranges, great bushes crowded with fruit, and in little cottage gardens bright sweet pink stocks, with such a smell of summer, and a yellow butterfly to make us almost believe we have found it already. As the sun set, the colours on the water grew heavenly in their beauty, the castle on the Ile Marguerite glowed like fire, and as a little boat shot into the light, its brown sails seemed to burst into flame.

We had intended to go to the beautiful *new* Hôtel

Toulon.

Gonnet, but the omnibus deposited us at the wrong one, and as we only meant to stay one night, we did not think it worth making a fuss about, and so settled to remain. Our room at the back is very comfortable, but the house is a queer one, and the company very second-rate and odd, the cookery poor, and the *salle à manger* a dismal little room, the paper torn away and the floor very dirty.

We left our good quarters at Marseilles at 8.30, and had a pleasant journey to Toulon. Securing our luggage at the station, we walked down into the town in blinding sunshine; we could not see the *forçats*, much to our disappointment, though it must be a piteous sight. A commissionnaire told us afterwards he could have obtained an order of admission for us to the Government works, but it was then too late. Toulon is a picturesque old town— steep narrow streets leading to the quays, roughly paved with the little round stones, which are so exceedingly fatiguing to the feet, and with water running through most of them; here and there a row of trees, or a handsome old building, breaking the uniform line of small shops and cafés.

We were much amused by our walk on the quays, and the people there, soldiers, sailors, peasants, townspeople, a party of dusty, tired-looking conscripts, men-of-war's men, sturdy handsome fellows, very much like the English, with their bright trim dresses and jaunty hats, which differ from those worn by our men; they are stiff and turned up all round, and must hold a great amount of water, and have to be baled out in a storm.

We lunched at Toulon in an old-fashioned inn, sitting in a long, low room, picturesque enough, with polished wooden beams and floor and wainscoted walls, leaning sometimes out of the latticed windows to watch a little group of market-women bargaining and dozing by turns under the trees, and a man asleep in the great empty omnibus at the door. We grew lazy too, in the warmth and drowsy quiet of the old room, but finding the straight-backed chairs, though in good keeping with the oaken beams and the polishing, anything but comfortable, we began to long for a summons to start, and to wonder why every moment seemed longer than the last. Our bones were aching, we felt fidgets down to the ends of our fingers, and we had a good halfhour yet to get through before the man in the blouse outside would wake up, and the horses be led out and harnessed.

My father took another survey of Toulon, visiting those parts adjacent to our house; but our boots were thin, and our weary feet, already blistered by the little pebbly stones, refused to encounter any more *pavé*. We had exhausted our resources of amusement; tatting and our invaluable book of 'Double Acrostics' had ceased to charm us, and we were so entirely *abîmé* with the heat, that we had no 'inner consciousness' to retire into. We were growing very cross, when a worthy bourgeois family entered, and effected a diversion.

There were two old ladies and two young ones, and one young man. One mother and daughter had invited the other two ladies to a gorgeous repast at the hotel, and, to make the fête complete, the *beau jeune homme* had been attached to the party. He was an ineffably stupid youth; the women—like all French women—made themselves exceedingly happy, and were gay and good-tempered, and the two young ladies, who sat on either side of their victim, made love to him and talked at him and across him, and pressed him to make himself at home, while he grew paler and more miserable every moment. We could not feel sure whether he was shy or dyspeptic, but he was a dreadful failure, and it required heroic efforts on the part of the joyous hostess to dispel the solemnising effects of such a skeleton at the feast. Sympathy for the bright-faced girls, and the hearty enjoyment of mesdames the mothers in the good things provided, made us forget our own discomforts. The long-tailed horses were being led out, and we mounted to our places, catching a last glimpse of '*ce pauvre monsieur*,' perched perilously near the edge of his chair, and gazing feebly at the demoiselles over a huge slice of melon.

Cannes, Tuesday.

This morning C. and I were out at seven o'clock, walking first through the little town, thronged with busy country-people holding their market under the trees. I made several sketches, and then, as we began to feel hungry, we turned into a clean-looking *laiterie* in the long street, and asked for cups of milk which were quickly supplied, the kind little woman sending to the baker's for three *petits pains*, and we sat down in the little shop on straw chairs, Madame considerately supplying us with two more for our feet, and a fifth for the bread which the man laid out in state on a piece of cream-white paper. It was a deliciously fresh dairy to breakfast in, and we enjoyed watching the customers come in with their cans and basins for the sweet new milk, which was poured out with many a pleasant interchange of morning greetings.

Madame brought us two serviettes, which she insisted on spreading over our knees, evidently thinking we were not to be trusted to look after ourselves, and assuring us that if any of the milk was spilled it would prove most injurious to our dresses.

'*Mais, Mademoiselle, cela fera beaucoup de mal, beaucoup, beaucoup!*'

We had a little gossip with her about her business and her customers, and the English, for whom she professed great regard; and she brought a pot of cream and poured some into our cups, explaining, '*Cela va mieux; c'est bien bonne, la crême, n'est-ce pas?*' as we tasted it, '*c'est pour vous, Mademoiselle, il n'y a rien à payer.*'

The cream being a *cadeau*, we paid fivepence for the breakfast. The honest little woman proposed that the *petit pain* which we had not eaten should be returned to the baker! We gave her two sous for the man who had fetched the bread, and she cried out, '*Ah ciel!*' in amazement at such generosity. Evidently the resident English in Cannes have not yet succeeded in destroying the simplicity of the natives. We parted quite affectionately with Madame, who was delighted to have one of our little French gospels.

Much refreshed by our Arcadian repast, we wandered on beyond the hotel again to the extreme end of the curve facing the islands, passing many charming houses, mostly quite new, with gardens full of orange trees and bright flowers, till we found some rocks and sand running out into the sea, and there we camped out for nearly an hour, luxuriating in the wet seaweed, and shells and pebbles, and the seaside smell, after our dusty railway journeying and long hot days in cities. The sea grew the most wonderful blue I ever saw as the sun rose higher, and the sky was exquisite beyond all words, the Estrelles hazy in their violet glow in the distance, and the palm trees and cypresses breaking up the foreground. As we walked leisurely back to our hotel, we watched a large boat full of English putting off, and then two pretty French girls bathing, who ran down the sand hand in hand. You can fancy how pleasant this road must be, as the water comes in almost close to its edge, just a narrow strip of sand between, with here and there a bathing machine or a little hut for dressing in.

We returned in time to join my father at a table d'hôte déjeuner, meeting the same dismal provincial French as on the previous evening, and one very pleasant Englishman. The meal was served in a larger and better room,

Market at Cannes

and the hotel also boasts a drawing-room, and a pianoforte for the amusement of the ladies who are here *en pension* for the season. There are very few carriages on hire at Cannes, even including those belonging to the hotels, and, as we failed to secure one in time, we could not explore the higher roads, and only saw Lord Brougham's villa, and other pretty English-looking houses and gardens, from the railway.

Nice: Tuesday, April 11.

We left Cannes at 2.5, and reached the end of our day's journey in good time in the afternoon. This place is wonderfully increasing; they are building in all directions, and laying out new streets and boulevards. We are at a charming new hotel, started by a company, with an Englishwoman as superintendent, and English and French servants—'Hôtel des Anglais'—still very full, though all the others seem emptying fast. A pleasant Scotch gentleman, our travelling companion, recommended it strongly, and as crowds of people who have been wintering in Nice are now hurrying homewards, we hoped to have been welcomed with open arms, and to find the best rooms at our service! But they made rather a favour of taking us in at all, one or two Russian families occupying the whole of the first floor. C. and I have a very pretty room *au troisième*. My father has the only single room vacant below. He still suffers a good deal from the rheumatic enemy in his leg, and the necessity of climbing many stairs makes him uncomfortable. It is all so fresh and clean that one can put up with the height. We are all very well, and my father looks ten years younger than when we left England, and is capitally sunburnt. C. and I do not suffer much from the heat, though in our longer journeys we have sometimes found it oppressive: the nights are still cool. At Cannes they were very cold, and the bedrooms were draughty. I think the warm weather must have begun all over Europe at the same time. They say here it was cold till the beginning of April, and in Paris the sunshine began on the same day as in England. The biting cold winds we had through March, even in Devonshire, have made us enjoy all the more this our first southern spring. All through our long railway days I had a pleasant sense that we were going south to find the summer, and every little flower and budding tree we passed seemed pointing towards it. I am glad you are sharing in the bright warm days; but we like, too, to consider England as under a dominion of east wind, and think M. mentions too ostentatiously that they have flowers in the open air!

We have no difficulty in our packing, and the weather is too fine for us to have missed our waterproof cloaks. My father is gone to the Post for our letters. I cannot tell you how welcome they all are.

Evening.

After E. wrote the above, we joined the very small table d'hôte, our Scotch friend sitting by us, and amusing me with riddles. I think our powers are well matched, and we lived through the evening under quite a shower of conundrums, having our tea together in the salon, and propounding sentimental poems over the sugar basin. There was no private sitting-room vacant, but the salon is very quiet, and when we are once upstairs, our pretty room answers every purpose, as we have a sofa, and tables, and easy chairs.

Cannes.

Cannes. 6.30. A.M.

IV.

Nice: Wednesday, April 12.

THE weather is still beautiful, though this afternoon has been rather cloudy, and the wind sufficiently cold to make us wish for more wraps when driving. After breakfast this morning, we started to explore a little, and had a charming walk and a very amusing one; examining the beautiful shops, many of them filled with coral in all shapes and forms, reminding us of those at Baden; passing through the flower market and buying bunches of deliciously-scented double violets, and longing to carry away the great bouquets of camellias and other lovely flowers. One of the tiny green frogs was hopping about at our feet, and we were glad to rescue it from probable harm, and restore it to one of the great market crates full of cabbages, in which it had no doubt journeyed, much to its astonishment, in the morning. We notice the single roses in full bloom in the garden outside our hotel, and at Cannes there were scarlet geraniums quite drooping in the heat!

When we were tired of walking, my father engaged one of the capital two-horse carriages, which are so common and so convenient here, and we drove to Villa Franca by the new road, passing many pretty villas, and noticing especially Colonel Smith's 'Folly;' he is the same who built on the sands at Paignton, and he has another of these strange erections at Rome. This one has been many years in building, and is still uncompleted; Moorish in style, with walls, and arcades, and mock forts serving as boathouses, covering the hillside to the water's edge. Colonel Smith tires of them, I believe, almost before they are finished, and has therefore less excuse for introducing such disfiguring objects into charming places. The views as we drove on were exquisite; the sea of an indescribable blue, dotted with white sails, and behind us the graceful outline of the bay, and the white houses of the town following the curve of the shore. In the Bay of Villa Franca, below us, were three Russian men-of-war, stationed there during the residence of the Empress at Nice. We left the carriage, and walked about the queer little town, climbing up the steep narrow streets, mostly a succession of steps, and noticing the public buildings, all on a most tiny scale—an hôtel de ville like a doll's house, with one window by the door.

On returning to Nice, we drove to the Hôtel des Étrangers, to ask about a vetturino for the Corniche, and then to the Hôtel de Nice, a new and handsome house built some way back from the sea, which is a great recommendation to many who do not like to be very close to the water. Having paid our call, we returned to our own hotel for luncheon, and in the afternoon took another drive through a valley with fine hills at each side, and the bed of a torrent or river (the Paglione), almost entirely empty of water now, with washerwomen drying their clothes on the stones; it must be a very considerable stream after heavy rains or the melting of the snow. We passed two large monasteries finely situated on heights, and the picturesque convent of Cimier far above us, but did not attempt to climb to them then, our destination being the grotto of Falicon, a picturesque cave lined with stalactites, at the foot of a hill shutting in the valley. We found, however, that a visit to it would necessitate a good deal of walking, and take more time than we had to spare, so we contented ourselves with driving into the court-yard of the old château, which is now to be let. The bell of the little chapel adjoining the house was clanging for the service, at which some priests we met were evidently going to officiate: its monotonous tones would hardly prove an additional attraction to possible tenants, and the place altogether looked gloomy and untempting. The air was so chilly in this narrow end of the valley that we were glad to turn our faces homewards. There is not much of beauty in the drive, but the fruit blossom in the cottage gardens was lovely, and oranges and occasional palms threw a little of the Southern charm over it all.

At the table d'hôte to-day were two French ladies who interested us; the younger one was very charming-looking, with glorious eyes and lashes, and wonderfully fair complexion, set off to advantage by a quantity of very light-coloured coral; her husband is a Russian, and she has wintered here for her health, which is very delicate. She is the most perfect work of art. Pencilled eyebrows, soft colouring like a blush rose, the best specimen of 'painting on enamel' I have ever seen. She would be an interesting study if it were not so sad a one, as the unfortunate lady is in a consumption. We asked our femme de chambre what was her name, expatiating on her loveliness. The girl cried out—'Ah, does Mademoiselle think her beautiful? As for me, I find her quite plain, but then, I have seen her in the morning. If Mademoiselle could only imagine her before her toilette is complete, the poor Madame—it is too horrible—she is pale and wasted like a ghost.'

This hotel is almost ludicrously English, from the manager and head-waiter, down to the little cans with 'hot water' on them left at your door in the morning, but it is certainly none the less comfortable on that account.

Thursday evening, April 13.

I am sitting in our room after the table d'hôte, and must try to tell you of our day's experiences. In the morning my father and I visited the Exposition here. It is on a small scale, of course, but quite worthy a visit, being full of native products and manufactures, amongst which the scents, preserved fruits, and wood work, both carved and inlaid, take a prominent place. There were some very beautiful billiard tables, and other pieces of furniture, with groups of flowers and figures in the different coloured woods.

We noticed a lady, with two attendants, who was examining the photographs and other things; and were exceedingly startled on hearing her speak, her voice having a curious spasmodic effect, like the bark of a dog. We have since been told that she is a Russian countess, whose story is very sad and strange. A serf on the estate of a great noble, she lost her voice in the terror of some sudden attack of her master's dogs. Her life was saved, but all power of *human* speech was gone. The count, in a sudden access of remorse or pity, married her, giving her the shelter of his name, and sending her here with a handsome provision for her life. This must have been many years ago, as she is not at all young now.

At one o'clock we went on board the little steamer for Monaco. The chief attractions appear to be the casino and gaming-tables lately opened there. The sovereign of this smallest of European monarchies lives in Paris, and he has sold the reversion of his principality to the Emperor, so that one may hope this gambling establishment will not have long to flourish; at present people crowd to it from Nice, and everything is done to add to its popularity.

The sea was quite calm, and exquisitely blue; and we very much enjoyed our little voyage, passing Villa Franca and the Russian vessels, and beyond it Esa, a queer little walled town perched high up on a rock, with the Corniche road winding above it. Monaco is built on the top of a hill also, and is very picturesque; but there is not much to see in the town itself. We found a good hotel connected with the casino, where we lunched; and then wandered about through

A little breeze on the Promenade Anglaise. Nice.

Spanish Pedlars.

the lovely gardens, full of bright summer flowers, roses
and salvias and lilies, and Banksia roses climbing amongst
such a beautiful Cobæa covered with blossoms, and a
real *orchard* of lemon trees in full bloom, and palms—
the sea a wonderful colour below us, with Mentone
away in the distance. We walked through the rooms,
and lingered for a minute here and there, to watch
some of the faces; but there did not seem to be much
high play. After about two hours we drove down
to the boat again, having rather more passengers this
time—some curious ones, full of talk about their good
or bad fortune and their systems of play; one dismal-
looking Frenchman amused me by the strange combination
of a parasol and spurs! We came back soon after six,
in time to join the table d'hôte.

April 14, Good Friday.

A very disagreeable change in the weather; clouds,
east wind, and the dust blowing wildly about the roads.
It was difficult to fancy that we were not in England,
when the waiter brought us hot cross buns with our
breakfast.

E. is taking care of a troublesome cold, which has
made her feel really ill, hoping to lose it all the sooner for
nursing now. She has just had her luncheon, for which I
ordered some 'strong soup,' to be *very* good; endeavouring to impress on the waiter that Mademoiselle required
something nourishing. So presently it arrived, and we
have been laughing over it and regretting the reckless
expenditure of vermicelli—a breakfast-cupful to a few
spoonfuls of the weakest and whitest of broth; and poor
E. was feeling quite an appetite, and looking forward with
enjoyment to her good meal. My father has brought her
a great branch of orange blossoms, for which he gave a
penny! a strange contrast, this sign of southern warmth
and sunshine to the keen spring wind blowing outside.

Sunday evening, April 16.

C. and I are sitting in our room, after the table d'hôte; and this letter must be only a short one, as we want it to go off to you as soon after the last as possible, and I have very little to write about.

We have had a very fine day, bright sunshine, but a terrible NE. wind. We walked after breakfast to the new and handsome English church near this hotel. It was crowded with people; the singing, congregational chiefly, was very lovely; and we had a most earnest and beautiful sermon from the rector: an Easter Sunday sermon, on the text, 'I am the Resurrection and the Life.' It was sad to hear many weary coughs, and to see the shadows lengthened on many a fair young face; one fine handsome man, sitting in front of us, looked fearfully ill; and I could not help wondering how many of those now gathered under that roof would live to meet another Easter Sunday. In the graveyard beautiful red roses were clustering round the stones—memorials of so many who had died in their early bloom. It is touching to see amongst so many tombs that almost all are for the young. . . . . .

V.

San Remo: April 18.

. . . We must tell you of our two days' experience on the Corniche. We did not leave Nice till about twelve o'clock, and so had time for a last walk after breakfast. The charges at the hotel were all very moderate, and we were so exceedingly comfortable there that it was quite hard work to leave it. Civilisation is very spoiling, and we are rather dreading the possible dirt and oily discomforts of wayside Italian inns.

It is almost ludicrous to see how English people carry their home comforts and luxuries with them, like true believers who travel with their own pottery lest they should eat out of the same dish as an infidel. So throughout Europe there are *restaurants* and shops for Bass's ale, tins of English biscuits, potted meats innocent of garlic, essences carefully preserved in air-tight cases to strengthen the soups, and at the 'Hôtel des Anglais,' Nice, we had hot cross buns on Good Friday; were called at eight o'clock by the chambermaid, who brought the hot water; and waited on at dinner by an invaluable butler, who one felt must have acquired his education in the very first society of his native land. That man alone would make the fortune of an hotel company; to watch how he made the tea in the morning, brought fresh hot toast, and placed the latest 'Times,' ready cut and folded, by your side, carved the ham for you, or selected the tenderest bits of cold chicken, was to behold a marvel of British civilisation; just a shade of solemn depression in his manner, as though it cost him an effort to be equal to the occasion, all the circumstances considered.

I like to travel with every comfort, or to make up one's mind to rough it thoroughly. In the mountains, one enjoys living as the people do, barring the black bread; a sort of knapsack journey has its own indescribable delights, when mind and body are braced and invigorated; and on the top of a mountain, after a hard climb, a crust of dry bread out of a guide's pocket has seemed the most delicious morsel one ever tasted. There is no sauce so piquant as hunger, but then in Switzerland one has very seldom to encounter dirt. In Italy there is very little cleanliness, and a great deal of oil, and I wish gratefully to chronicle my sense of Huntley and Palmer's enterprise, and the value of many a little tin case sent forth from Piccadilly.

No doubt real Italian hotels are rapidly improving, and it is tiresome to have too English a stamp on everything on one's road, but in towns out of the way of the general rush of travellers, they are still wonderfully behind the rest of the world. Do you remember my father's story—how, when in South Italy they ordered hot water to make some tea (having a supply of the same with them), and a little was brought in a tumbler, into which the tea was put, when seeing their perplexity as to how it was to be drunk when made, the waiter seized a wet towel from its stand in the bedroom, and, reversing the glass over it, squeezed the water through into a cup, which he presented triumphantly for breakfast.

But this is a long digression, and I must return to my journalising. We made a capital start, our carriage being roomy and comfortable, the horses good, and our driver very good-tempered, though not especially communicative or interesting. Just outside the town we found another horse waiting to take us up the long steep hill, which wound in wonderful curves and zigzags, amongst villas and gay gardens, and then past orchards of various kinds of fruit trees, and hedges of aloes, with lovely views over Nice and the sea, and the Estrelles in the distance, and nearer to us a valley and mountain stream far below. The man belonging to the additional horse amused us by his great activity and apparent insensibility to fatigue, as wherever the road was at all level, he urged the horses on with whip and voice, keeping up with them all the time, and not seeming to mind the thick clouds of dust that swept over him continually, and which, with the great heat, hardly suffered us to breathe. Literally the men *belong* to the animals, though *they* may possibly owe some allegiance to a master *voiturier*; but anyone who has seen the women in Holland or Germany, and indeed throughout Europe—their whole lives devoted to watching

a cow, moving when it chooses to move, and standing while it feeds—will understand what I mean.

Another vetturino kept in front of us, and we were constantly meeting others, and passing groups of peasants, mules with gay southern trappings, and children with flowers for sale, one of whom we sketched, to her great amusement. Soon we could look down on the lovely little bay of Villa Franca, with its deep green water, and the Russian vessels lying at anchor; and every turn of the road gave us some fresh glimpse of the beautiful coast, bringing us at length within sight of Esa, a little walled town, perched on the top of a rock between two valleys, and looking down on the sea. Monaco lay far below us, with its new buildings and old walls, and independent government—as Lilliput must have looked to Gulliver—resting on the soft blue Mediterranean, which was showing us its fairest colours, changing and shimmering under the light, fading into a soft haze on the horizon, and deepening into glorious purple and green under the shadows of the rocks, and iridescent where the tiny waves caught the sunshine. Then Turbia, with its old Roman remains, came in sight, the road passing through the queer little town; and so, by a pleasant descent always close to the sea, we came to Mentone, which is very different from Nice or Cannes, with its one narrow street full of people (Easter holiday-makers thronging in from the country), growing at either end into a pleasant boulevard, with villas in gardens, and houses, large or small, *à louer*, bordering the road. High wooded hills rise behind them, and from the back windows of the opposite houses you look out upon the sea. The chief hotels are on the further side of the town.

We drove through an avenue of plane trees, and stopped at the Victoria, which is close to the water and at the beginning of the main street, and more amusing for one night, though we were not especially charmed by the accommodation or fare. The sea was quite rough, and the great waves came in grandly; we walked close to it and through the little town, looking at the many good shops, pretty wood-work, and picturesque people. The whole place is surrounded by hundreds and thousands of lemon trees, in full bloom. We walked through a garden rich with their heavily laden boughs, the sun shining down through the bright green and yellow leaves on the long fresh grass, every branch covered with ripe golden fruit. The lemon *masses* much better than the orange, which looks stiff when planted thickly in groups; but these trees were like an old English apple orchard, the fruit and leaves being changed, and the air scented with the perfume of the flowers. There is something very charming in realising that we are at last face to face with this prodigal southern nature, so doubly beautiful in its fresh green life, after the cold winds we have left in England.

We breakfasted earlier than usual, and, leaving my father to indulge in a longer nap, walked up and down the little street, Rue St. Michel, the principal one of the town, visiting Amarante's bazaar (a shop of inlaid woods, and general depository) and the market, with the peasants crowding on the old stone step and wall that marks off their territory from the road, and hunting up some washerwomen in a queer back yard or garden, who let me draw them as they knelt in baskets perched round a great tank, and who were delighted at receiving some Italian gospels, which I fancy they considered as lives of the Saints. One old woman seized on a San Giovanni and reverently kissed it, and we find 'San Marco' is very popular.

We sat some time on the broad platform above the sand, watching the great emerald-coloured waves breaking

in foam at our feet, strolling on a little way by the water's edge, then turning into the street again, which runs parallel to the sea; the large houses facing the water at the back, and with their front doors opening on the road. Mules with red and blue tassels and every kind of rope harness, women and men in gay dresses, the former with bright-coloured handkerchiefs twisted round their heads, the men in curious pouch-like caps, brown or crimson, green shirts, and scarlet scarfs round the waists, or high conical hats and long picturesque beards; a tall handsome race. The women are many of them really beautiful, with such grand features and outline, glorious eyes and hair, and often very good complexions, clear, with a brilliant colour; they are well grown, and even the very old women show some remnants of beauty in their stern massive faces. They all carry fearfully heavy weights on the head; we saw four or five girls come in to Mentone this morning with baskets of lemons, their heads perfectly steady under the burden, but the whole body seeming to move from the hips as if dislocated. They have slender ankles and feet, often bare, and wonderful taste in arranging the brilliant colours of their costumes. The avenue at Mentone looked really green and almost shady this morning, the plane trees all bursting into leaf, and people were starting off in their summer dresses for drives or donkey rides up into the hills, whilst little English children and their nurses strolled about, their hats covered with white muslin to keep off the sun and dust.

The town is well sheltered from off-shore winds by the mountains, thickly clothed with olives, which rise abruptly behind it. Every little piece of rock has been cultivated by the industry of the inhabitants, who find their toil well rewarded by the rapid growth of orange and lemon trees. It is in the latter that the chief trade of the little town consists. The olive tree grows slowly; many of the largest along the Riviera being hundreds of years old. Their crops are uncertain, being liable to failure from many causes, especially too great drought or violent rains. They always need constant care, and a considerable amount of labour to be yearly expended on them. The roots require plenty of air, which necessitates diligent digging; and in the early spring they are well manured with rags, an expensive luxury to the poorer proprietors, who have no extra clothing to spare, and have to invest in the supplies sent mostly from Naples. The leaves are very subject to blight, which would seriously affect the crop if not cured.

I should infinitely prefer Mentone to Nice as a winter residence; it is thoroughly Italian still, which Nice is not. For a great invalid, requiring home comforts, the latter town must offer more attractions; it is very much like a fashionable English watering place—Brighton or Torquay, with perhaps less available good inland scenery than the latter place. The air is said to be very stimulating; I cannot judge from my own sensations, as I felt ill there from my heavy cold; but after the bitter wind which swept in great gusts over the place, making one fancy, even in the side valleys, that winter had come back again, and blinding us with dust and sand, Mentone was a very pleasant contrast in its little sheltered nook, nestled under the wooded hills which offer endless charms to the visitor. I do not know that it has a club-room or a library; but it offers a bright, fresh contrast to English life, while it possesses every comfort and necessary for quiet country enjoyment. When C. W. was there last year, she found a fresh walk for every day of the month they spent there; and glorious indeed must be the views of the Mediterranean seen through ever-fresh settings of greenery. *Chacun à son goût.* There are people who would prefer the Promenade des Anglais, Nice, where you can exchange bows with all your acquaintances, and have an

uninterrupted view of the water; I am only speaking of Nice proper. There are many lovely villas in its environs, and a more beautiful distance to be explored; but such a residence necessitates a large expenditure, a carriage and staff of servants; whereas, at Mentone, lodgings near the sea, or apartments in a good hotel, and a donkey, make you free of all the enjoyments and charm of the place. Forgive this long prosing, but I like to have a clear idea in my own mind of the relative attractions of Nice and Mentone as winter residences, and certainly, if we should ever try either of them, C. and I should choose the latter.

We started at twelve, sorry to leave this primitive pleasant little place. My father laid in a store of fresh bread, and our other provisions we had bought in Nice. A long ascent led us to the French, and then to the Italian frontier. At the former we watched the official with a long fine steel rod, like a knitting needle, two yards long, and a face and eyes something like a ferret, delicately probing the fat sacks that surmounted a cart full of wood; the sort of man who can not only see through a stone wall, but take quite a wide look round on the other side! At the Italian Custom-house we failed to impress five or six *douaniers* with a due sense of our respectability, for they actually took the trouble to take down one of our boxes, and I believe even to open it; but it was the affair of a moment, and then we were off again, and able to begin our luncheon.

The Alpine Club has been so laughed at for expressing its appreciation of bread and cheese and beer, when a refined critic was waiting for a burst of enthusiasm over the scenery, that I am half afraid to trouble you with the fact that we *do* eat, or to tell you what we live upon; but as we promised minute particulars of our journey, and I never heard of any traveller but Lord Byron who was ashamed of owning to being very often hungry, I shall tell you boldly that we are *always* so, and that our little pic-nics will be amongst the most amusing and pleasant recollections of our drive along the Corniche. But to return to facts: C. and I were very hungry, and she had been desperately afraid that the officer would declare our pâté de Strasbourg illegal, and suppress it. We had a bottle of Bass's ale in one corner of the carriage, and my father had drawn and replaced the cork. We were quietly ascending a hill, C. and I watching with anxiety for the moment when our food would be distributed, and all hands full of hunches of bread, and our laps with gloves, books, salt, knives, &c., when there was a sound as of the breaking of a dam, followed by a rushing torrent, which, on looking round, I saw foaming behind me, and we were literally flooded by the beer. Were I 'Umbra,' I might here state that it rapidly rose to our knees, and that C. was drowned before my father and I could extricate her from the carriage! But the unvarnished truth is that I was soaked; that C. held all the pieces of bread and bits of pâté, and knives and gloves; that with great presence of mind I put the salt in my pocket and sprang into the road, while the *cocher* philosophically mopped up the deluge with an old rag, and then turned the cushions by way of cleaning them! I was rather damp and inclined to be cross, but I remembered in time having heard W. or the dressmaker declare that washing in beer was good for silk dresses, so I recovered my equanimity and subsided into a state of dignified calm. Notwithstanding this misadventure, our luncheon was a great success.

We passed many mountain streams; at Ventimiglia the broad bed of the river was almost dry, except in places where men were busy constructing a new bridge for the railway, working up to their waists in water; the valley looked wonderfully picturesque, with the grand old brown dilapidated painted town piled up against the hills on one hand, masses of olive trees and great palms clothing their sides, and groves of lemons running round the base, range upon range of hills soft blue and violet closing up one end, with the higher mountains crowned with snow. The stony bed of the river was gay with people, men working, women washing, idlers on the long bridge, and at the other side the green Mediterranean foaming on the beach; these lateral valleys form one of the most interesting features of the coast.

We were interested in watching the progress of the new railway, though it was rather a sad sight. One does not like to think of this road ever being deserted, and the little towns and inns falling back upon themselves for life and employment. Anything that advances the real prosperity of a country must be good in the end, and the few must be content to suffer for the well-being of the many. But I cannot help regretting it from the traveller's point of view; we English are far too prone already to get through the greatest possible space in the smallest possible time. We saw no workmen on the line, except about the bed of the river; by the roadside there were stacks of rails waiting to be laid down, and the bridges are almost all finished, and many of the tunnels. The line is a single one, and runs mostly between the road and the sea, almost touching and often crossing the former. For miles it is carried along the beach, so close to the water that you might play ducks and drakes, if you were quick enough, from the windows.

The landlord at Mentone told my father that the lemons are sold at three napoleons the thousand, which strikes us as dear. It is curious to realise that all the 'dried fruit' we eat here is 'grown on the premises;' the little figs and the almonds are very good, but the oranges are thick-skinned and sour. This evening at dinner, with some delicious *fresh* sardines, they brought us lemons just gathered, which scented the whole room with quite a wonderful perfume. The air was cool to-day, a good deal of wind, but some south in it, and we were able to have the carriage open, and did not suffer from dust, though it lay thick on the road. We have had a charming drive, passing groves of magnificent palms as we drew near San Remo; we are re-reading 'Dr. Antonio,' and grew quite excited over Bordighiera, realising each turn of the road, and trying to fit it to our book, and arranging which of the little roadside houses could be *the* Osteria. It was amusing to see the different styles of architecture (all supplied by the painter) in the houses, many having entire false fronts facing the road—jalousies, balconies, and handsome copings in high relief and wonderful perspective. Here and there were tumble-down old villas, half hidden in masses of lemon and orange trees and peach blossom, with scarlet geraniums and sweet-scented stocks peeping between the stone pillars on the walls, on which the supports for the vines will rest in the summer; the loggia rising grandly over the green leaves and the painted old walls, with its roof of bright, irregularly-coloured tiles, and a great palm or a cypress towering beside it. A grand palatial entrance in the worst style of art and stucco, and a broken-down door.

We drove through San Remo to this hotel (d'Angleterre), and were received with a rapturous smile by the landlord, who, on our enquiring as to the possibility of beds, assured us there were plenty; and on our alluding to a table d'hôte, told us that we could dine at any hour, as at present there was no one in the house! So, with the pleasant sense of performing a good action by the mere fact of coming there, we took possession of comfortable rooms looking over a garden gay with scented stocks, and beyond it the high road passing under an avenue of plane trees, which we have almost watched budding, under a soft spring rain that has been coming down in good earnest since our arrival; and we have been sitting at our open windows, enjoying the cool plashing sound of the great drops on the thirsty ground, and watching the holes they make in the dust.

This dawning of fresh, green life is very beautiful, every little bough and twig is wakening from its long winter sleep, and there is everywhere a stir and whisper of coming spring. Maurice de Guérin, in some of his pictures of nature, paints, with that wonderful insight which is the great charm of his writings, this same season: 'The woods have not yet got their leaves, but they are assuming an indescribable air of life and gaiety, which gives them quite a new physiognomy. Everything is getting ready for the great festival of Nature.' And later: 'One can actually *see* the progress of the green; it has made a start from the garden to the shrubberies; it is getting the upper hand all along the lake; it leaps, as it were, from tree to tree, from thicket to thicket, in the fields and on the hill-sides; and I can see that it has already gained the forest, and is beginning to spread itself over its broad expanse. Soon it will have overrun everything as far as the eye can reach; and all those wide spaces between me and the horizon will be surging and moaning like a vast ocean—one sea of emerald.'

We passed some acquaintances on their way back from Genoa; and while we were at dinner an English family arrived with a great noise, four horses charging at the hotel; and a quiet gentleman travelling alone, with whom my father has since been walking in the town. I am feeling much better; the cooler air has been so pleasant.

The beautiful palm trees give the great charm to this place; but one could spend days exploring the neighbourhood of these towns on the Riviera, each a centre of interest, local and traditional, of its own. It was from here that the brave sailor came, who, standing among the crowd in front of St. Peter's, watched the mighty obelisk slowly rising to its place; and who, when the people, forbidden to speak on pain of death, saw with an agony of dread the slackening of the ropes, cried out '*Acqua, acqua!*' saving it from destruction, and winning pardon and applause from the grateful Pontiff, who bestowed upon him and his heirs for ever the right of providing the Easter palms for Rome. The branches each spring are tied together on the trees to bleach them sufficiently, and thus their beauty is a good deal lost; but we saw beside the road many a grand old tree with its broad feathery crown, each bough still waving gracefully in the breeze.

Alassio: Wednesday, 19th, 1 o'clock.

While we are waiting for our luncheon, I may as well continue my journalising.

We have had a glorious drive of twenty-nine miles from San Remo. The rain which fell during the night was just sufficient to lay the dust; and we opened our eyes on a perfectly fine morning; breakfasted at 7.30, and took leave of our very attentive host. The carriage was open, and my father mounted the box, leaving us the whole inside, where C. and I established ourselves most comfortably, our bags and loose parcels ranged on the opposite seat. We started at a good pace, which the horses kept up all the way without any great effort.

I am writing in an enormous, gaunt room, nearly thirty feet high, with two rows of windows, six doors, and a stone floor; the walls hung with a wonderful collection of oil-paintings and old prints. Our waiter speaks only Italian, and we have been laughing over our vain attempts to get any food we can eat. Some mutton cutlets redolent of oil, drove my father to his phrase-book; and the waiter, eager to please, took away the obnoxious compound, returning with some fresh roast meat and potatoes, which he assured us were cooked in butter. This was something *very young*; white meat, soft and tender; it might have been 'baby,' but C. thinks it was very little

Baby show at Alassio.

pig! Next came three small birds with the most piteous expression in their eyes, their long bills turned down over the poor thin ribs with the most sentimental action, their legs crossed, and arms akimbo, looking like very withered old men. We each dissected one, and then tried to swallow the omelette, but that was hopeless; so instead of eating I am writing to you.

In many places during our drive yesterday and to-day we passed hedges of luxuriant red and pink China roses in full bloom, growing half wild. We watched the railroad with great interest; it is wonderful to see over what places they carry it; but just now they are in a difficulty for want of funds. It is a single line, and, if it is ever finished, I think the traffic will often be interrupted; as it must be very liable to accidents from falling stones from the rocks above, landslips, or storms from the sea.

We passed through several very picturesque old towns bordering the sea, some of them looking really magnificent in the distance, with the houses piled one against the other, gaily painted, and with their cool white stone loggie, and the church towers shining against the blue sky. It is only when one is close to them that one sees the paint is rubbed off, and the stucco broken away. The streets in the smaller towns very narrow—so that it was difficult for carriages to pass between the houses—are very miserable and dirty, with heavy stone arches thrown across, and in some places story on story built above them. We passed by forests of olives, grand old trees, their soft twilight verdure resting the eye, wearied with brighter colour and beauty. In the grass beneath, women were hunting for the ripe fruit; by the roadside hedges of aloes bordered the fields, and peasants were ploughing with two oxen and the old wooden machine of the Romans. We passed villa after villa in the midst of gay gardens, peach trees in full blossom, fields of rye and barley in the ear, ripe peas and beans, little shrines for the Madonna, overgrown with Banksia roses, and festoons of vine and ivy. We stopped for two hours at Alassio, where I began to write, and, after our failure as to luncheon, went to the door of the inn opening into the narrow street, and there I sketched a mother and pretty dancing baby, seated on a doorstep. The child had a little cap and cotton jacket, and the very slight addition of a large white duster fastened round the waist, and flying open in the air, showing the fat baby legs to the greatest advantage as it jumped and crowed. The mother was delighted, and so many other mothers and babies offered as models, with such a large concourse of children and even men to look on, and all eager for our little books, that we made a hasty retreat to the beach, on the other side of the houses, where we were joined by my father. Here we found the fishermen busy with their boats; fine fellows, with their legs bare above the knees, and gay with bright-coloured scarfs and caps. I drew a good many of them walking up and down the sand, with an enthusiastic following of 'youths and maidens,' and their grandmothers, whose shouts of delight as they recognised a fresh friend or relation in my book grew so fast and furious that we were driven back again to the carriage, which we found just ready to start. A French family, in a large voiture à quatre chevaux, were before us, but we kept them in sight the whole way, and had a most charming drive of fifteen miles, often close to the sea, which lay like a calm mirror in the evening light, reaching in good time this place, Finale, where we are to spend the night.

We are established in a queer old Italian hotel, with a very civil landlord and very tolerable rooms, though, of course, rather dirty. The street is so narrow that we could talk across it! We are all well; this pleasant drive through the fresh air has quite set me up again. You can fancy how intensely we are enjoying this beautiful Corniche, and that we do not like to think how soon

Our dinner

Sketches on the Corniche.

Pescatore.

Charcoal carriers.

it will come to an end for us. We hope to reach Genoa to-morrow, and probably shall spend Sunday at Spezzia. The people all along this road strike us as singularly handsome, with good features and clear colouring, the nose finely shaped, and the face also, with a little droop about the mouth, and slightly hollow cheeks, the eyes fine, but often a little near together, the forehead low and broad. I notice two curiously distinct types: some of the boys are like models for Murillo, broad round faces, with glorious eyes, deep hollows below them, and very prominent cheek bones—a sort of Mongol or Kalmuck race. They have that *pale* colour that in a dark-skinned people looks yellow in the high lights and grey towards the shadows. I cannot account for this strange difference, which is very marked, as most of the Italians along the coast are of the purest Circassian type, the features almost perfect, and the colouring warm and bright. It is very strange to trace this marked variety in a race; they remind me in degree of the western and southern Irish, the long lantern-jawed, or short-faced and often pug-nosed peasant. Possibly on the Riviera there may have been in old days a striking difference between the inhabitants of the coast towns, with their mixed populations, and the mountaineers and contadini, which, now that they are more mingled together as one people, still shows itself in their descendants.

VI.

Genoa: April 20.

NOTWITHSTANDING the dirt at Finale, we managed to get some sleep, the iron bedsteads and bedding being tolerably clean, but a mouse that made a tour of investigation ended his peregrinations by running over C.'s pillow, who thereupon woke me to light a candle. There were dim holes and corners in the great room; old doors, half concealed by the paper of the wall, which was peeling off in strips that rustled uncomfortably; and we had caught a glimpse of a dark passage at the back of our beds, into which one or more of the doors opened. There was an uncanny smell and general sense of stuffiness, and once we were really sure a key turned in the lock; but we coughed significantly, and silence was restored. Altogether the candle *was* a comfort, and with its help we went to sleep again.

We decided to breakfast at Savona, so C. and I only had a glass of hot milk, and then we started, as before. The drive in the early morning was delicious, the air sweet and clear after the dirty oily inn and close street, which seemed a mixture of fever and garlic under our windows; and the sea was exquisite, with brilliant pearl tintings of green and lilac, and white lines of light, and a silver track for the sun—the hills and valleys all clear and fresh, and every bough and leaf on the trees cut sharply against the delicate colouring of the morning sky.

It grew very hot before we reached Savona at ten o'clock. This is a wonderfully picturesque old town, with a considerable amount of shipping; a great place for brick-making and potteries, and the men and women were busy in open yards by the roadside. I could not see any potter's wheels; the people seemed to be making a series of mud pies, with covers, which were ranged in rows in the sun, and turned somehow into dishes and saucers. We stopped at the Hôtel de la Poste, to give the horses a two hours' rest, ordered a déjeuner à la fourchette, and went down into the beautiful garden of the inn with our smiling old host, who gave us bouquets of lovely flowers. There were arcades of white and golden Banksia roses, hundreds and thousands of blossoms, lilac trees in full flower, lemon

trees scenting the air, red roses and pink China, and sweet-scented tea roses in abundance. The Mespilus Japonica grew there, but the fruit was not ripe. Two green parrots flew about among the orange blossoms, hardly to be distinguished from the bright-tinted leaves.

We walked down to the quays, and watched the people busy lading and unlading the boats, turning up afterwards into the narrowest streets it is possible to conceive, where the market was held and there were open shops filled with gay-coloured cotton goods. After our breakfast, which was very good and without oil, C. and I ran down again to the shore, and I made a hasty sketch of the market, and a high tower with a curious arched tunnel in it leading to another street. The people crowded round to look, and we gave one or two Testaments to them, which caused a general rush and eager demand for more, so that we were sorry not to have brought a larger supply from the carriage. I must not give you a false impression; the outcry for our little books came, I fancy, in ignorance of what they were, and C. maintains there would have been equal gratitude for sugar plums, not to mention sous! It may have been only the universal desire to share in any backshish that might come in their way, be it goods or money; but I fancy some good people have been hard at work about this Riviera, trying to raise and purify the creed of these poor peasants, and that they welcome the colporteurs, and gladly gather to hear their good tidings. The French Government encourages the sale of any books printed in that language throughout the newly annexed territory; and, as all the inhabitants of the Riviera are kinsmen still (as far as race is concerned), any new thought or feeling must rapidly spread. It is sad to see this fair arm of Italy bound to its new possessor. Nice might have been given up not unwillingly, and it is such a cosmopolitan town, that it could easily adapt itself to any rule, being half French in many

E

of its tastes and sympathies. The new order of things presses heavily upon the inhabitants of the towns along the Corniche. At Mentone, prices have been suddenly raised, and taxes are increased; *polenta*, on which the poor chiefly subsist, is taxed one sou the litra, and costs six sous in the town, while on the Italian side of the Douane it is to be bought for five—a state of things, of course, leading to incessant smuggling; the conscription, too, is most unpopular, as the poor peasants, no longer fighting for Italy, have not even a rag of enthusiasm to wrap themselves in.

We made a fresh start at twelve o'clock, and our drive for the next few hours was less interesting. Before reaching Savona, and about Noli, the road winds round the base of precipitous cliffs, which have been cut away to make it, and there are sheer precipices above and below, with the green waves dashing on the rocks, and here and there a picturesque fisherman to be seen with rod and line. The distant views were perfectly lovely, each fresh turn of the winding road bringing into sight new headlands and gracefully curved bays, dotted with white towns and towers along their edge. These Italian buildings are wonderfully beautiful in the distance, and, as 'Dr. Antonio' says, his countrymen have a natural taste and love of the picturesque, which makes them choose just the best possible site for every wall and building; but as you draw near, the spell is broken, and not even the ideal atmosphere can make you believe in paint and stucco and miserable decay and dirt being anything but its own reality.

After leaving Savona, the drive was less rich in beauty and the great heat made us all sleepy. We, who were afraid of being caught napping in the midst of so much to interest and charm eyes and mind, were rather comforted to see my father's head give a decided nod as he sat on the box, a sort of *side* nod that could in no way

A Genoese 'Pealer'.

Genoa.

be mistaken for a bow of acquiescence at any remark of the vetturino; he—worthy man!—smoked his pipe by special permission, and drove on steadily, with a lazy defiance of the railway and its power to force him off the road, in the very slouch of his broad shoulders.

At Voltri, about ten miles from Genoa, we came to the railway really in working order, and trains running; from there we passed through a number of small towns and suburbs, meeting hundreds of people driving out in the cool of the evening; open carriages filled with bourgeois, young and old women in their pretty white muslin veils; numbers of very young girls, looking gay and smiling, dressed all in white, with immense bouquets, going out into the country to make a fête day of their 'first communion.' Some had wreaths of flowers on their heads; and one small child we saw walking in the street, magnificently dressed like a Bambino, in white satin, with flowers and veil! C. and I were almost terrified at the narrow streets, and could hardly realise the possibility of breathing in such places. There is only just room for a carriage in many of the smaller ones; no side path; and foot passengers have to run where they can when a great omnibus, or *voiture*, with its jingling bells, comes trotting past. We were sorry to feel that our pleasant pic-nic life was at an end, and to think of parting with the comfortable, dusty, old carriage and driver. We have travelled leisurely enough thoroughly to enjoy all the varied charms of such a drive; and you can believe what good care my father takes of us, and how much he likes to recall old associations with the many places he knows so well.

Turning up a sort of blind alley, we stopped at the Albergo Reale, mounting no end of stairs to a very complete little set of rooms, in which we are feeling quite at home. The great *salle à manger* is on the first floor, with double rows of windows, and we are on a level with the upper ones, and have a charming sitting room and good airy bed rooms adjoining. Our salon windows look down on the queerest scene; a street so far beneath us, that it is almost invisible; a railway on one side of it, passing through the town, apparently regardless of carriages or people; a raised promenade on high arches, which separates the port from the town; beyond it, the port itself crowded with vessels, and the heights of Genoa crowned with forts. It is a most complete *vue en ballon*, and no words can describe the strange effect. My father was willing to submit to the climb for the sake of the purer air; and it is most amusing to watch the people and boats. Our bed-rooms fortunately look into a very quiet and always dark and cool street, about two yards across. We can neither see the sky nor the earth!

Friday.

This morning we started at ten, with a good *valet de place*; visiting first two or three churches, and the Duomo, where there were one or two fine paintings. I am not going to weary you with any description of the buildings. They are very grand, and fine specimens of different marbles, neatly joined; but then one can have that sort of thing in an inlaid table, or a hall floor, and I think there is more real beauty in a little Gothic chapel with its old stone carvings, than in the grandest of these more pretentious edifices. The outside of the Duomo, with its horizontal stripes of black and white, is the ugliest part; it does not tone down with time, and looks irreverently fresh and staring. There are two grand old lions at the entrance, and, what to me was the most beautiful part of the whole, some quaint figures standing on griffins, and resting against light spiral columns running up at the side of the building, but detached from it. The colouring inside was good, though the glow of rosy light was only gained through the use of some old red curtains. There was a wonderful mass of brilliant gilding about the roof, and on the capitals of the pillars, and some fine paintings in the spandrels of the arches. In almost all the churches were enormous bouquets of red and white camellias, left from Easter Sunday.

We walked up a steep street to Santa Maria di Carignano; there service was going on, and we waited and watched the people, and climbed to the roof for a view of the city, but were quickly driven down again by the intense heat of the sun, reflected from the stones and slates below us; after examining some good paintings, we walked back to the Via Nuovo, where we were to see the palaces. We could not visit the *Rosso*,—an immense building so named from its colouring,—as the Marquis is just dead, but we went through the Palavicini, which looked brilliantly fresh and gay with flowers; large plants of azaleas standing in the marble vestibule, and a cool green garden in the centre of the court. As we left, the old servant broke off large branches of the beautiful lilac blossoms and gave them to us, and they are now making our room look quite homish. I need not describe the different palaces—the Balbi and Durazzi—which we visited; their real magnificence striking us the more from our previous stucco impressions of Italian buildings. The rooms in the Durazzi were really princely; the hangings, carpets, gilding, all good and fresh; two beautiful silver vases by Cellini interested us, and there were some fine Van Dyck's, but an immense number of the pictures one does not care about. In the vestibule of this palace is a noble group in marble, by a Genoese artist, of the present Marchesa (née Palavicini) and her young son, an only child. It is very simple and beautiful. He is leaning on her knee, and looking up into her quiet, grave, sweet face; the hair and drapery, and the whole pose of the figure are charmingly arranged. The broad marble stairs and pillars make this palace very celebrated.

We came back to the hotel to a five o'clock table d'hôte. The Baron de C. and four ladies sat opposite to us. The youngest we decided must be Madame la Baronne, the others her sisters-in-law. Their pleasant-looking maids, and a footman in bright green livery all over, like the Savona parrot, were making a tour of the churches this morning. The ladies were very quiet, and we were shy of talking German across the table, and were all rather silent; it is so difficult to make 'table talk' amongst one's own party, unless it is a large one.

We like this hotel thoroughly, and there is a very attentive secretary speaking English, who looks like a clergyman or a professor from one of our Universities, overworked and out of health, and it is sad to see his meek and polite submission to the imperious little land-lady. He toils cheerfully up and down the steep flights of stairs, showing people to their rooms, and even answering our bell, and bringing soda-water on a tray, much to our mortification. This evening we drove for an hour, enjoying the cooler air; there are an immense number of *voitures* on hire, very good, with a pair of horses, for two francs an hour! In England it would be impossible to procure such a one without ordering it beforehand from an hotel, and then it would be very inferior and much more costly.

The narrow streets were crowded with people; one could almost have walked on their heads; and, as in all these Italian towns, the inhabitants swarm over the road, regardless of pavements where there are any, moving lazily aside as the frantic cries of the coachmen or the heads of the horses reach their ears. We drove round and round the public gardens, meeting many well-appointed carriages, with no end of Marquises from the different palaces, groups of soldiers and officers, and the Bersaglieri with their waving green feathers and jaunty hats. The garrison here has just been changed, and we watched a regiment of the line embarking this morning for Palermo. Many hundred men were drawn up in order on the great platform below our window; at a given signal they piled their muskets, took off and fastened up their blue coats,—some very fat, painstaking officers growing hot and angry in their efforts to teach the young soldiers how to fold them regulation size—and then they all swarmed down to the little pier in their light marching dress, like bands of white ants, each carrying a dark burden, and so were packed on board a floating sort of raft boat, as close as they could stand, and towed to the steamer for Sicily. How one rejoices to feel these two ends of Italy belonging to each other at last!

The fresh white muslin veil, *pezzotta*, is pretty enough on young girls or graceful women, but it looks dolefully out of keeping on poverty-stricken or wizened old ones creeping about the streets. Some of the very poor wear long calico scarfs over the head, with bright colours or patterns of flowers. At the palazzo Palavicini we saw more of the glorious bouquets of red and white camellias, hundreds of blooms in each. Those in the church were brought from the gardens of the villa belonging to the Marquis, which we are to visit to-morrow. Our drive this evening after the great heat was most refreshing. We passed by some gaily lighted gardens, a café where a good band was playing, and under a chestnut avenue, the trees in full leaf, where the lamps gleamed like fire-flies in the green shade. I must not forget to tell you of a street-side group we passed this morning; a number of men and women eagerly surrounding a queer-shaped gridiron over a pan of charcoal, cooking and eating green snails!

We have been sitting quietly in our room for the last two hours, my father and C. reading, while I write to you. We have secured the coupé and one seat in the banquette in the diligence for Monday. This will take us through in one day, which the *voituriers* refused to do; and we did not wish to reach Spezzia later than Monday night, or to travel on Sunday; and as C. and I have never been in a diligence, it will be rather amusing and of course infinitely cheaper. Our old *cocher*, who was charmed with the *bonne main* my father gave him, is very anxious to take us on; but then, he and all the others must have two days, even with four horses, and we do not want to sleep at Sestri, or lose another day.

Your letters have all been so welcome.

Saturday evening.

After breakfast this morning, we set out for a walk, visiting the coral shops in the Strada degli Orefici, a curious winding little street or alley, with its relic of the ancient goldsmiths' guild, a 'Holy Family,' a painting of considerable merit, framed and glazed, and looking down from the walls with an imaginary blessing on the profits of each little establishment below. If it be meritorious in the eyes of the Saints for the faithful to spoil the heretics, there must rest an extra weight of benedictions on the salesmen in some of those dark little cells. We amused ourselves with buying a brooch at one, a comb or bracelets at another, and comparing the prices with those we had been asked at Nice. In all these towns the coral is much the same price as in Naples. The filagree work was very pretty, but we saw nothing very new to tempt us. We followed the *valet de place*—through dark, gloomy, dirty arches, past little low shops, and amid laden mules and crowds of people, jostling against each other—to the fish market, which was a wonderful sight, though the places and the people were all so *filthy* that we needed hands and eyes to take care of our clothes. The fish were of every shape and colour, and it was well worth encountering the mob to see them. You know how interested my father is in the markets; and these back streets of a great city teach you something of the real life of the poor.

One is so apt to think one has seen a town thoroughly after having driven through the main thoroughfares and walked on a boulevard!

We started from the hotel at twelve, in an open carriage, for the Villa Pallavacini, but the clouds of dust drove us back, and we contented ourselves with visiting some charming gardens close to the city, belonging to a Genoese gentleman. A very intelligent gardener took us round or rather *up* them, terrace above terrace, beautifully laid out, with an endless succession of fountains and brilliant plants. My father was charmed by a new variety of camellia, from Brescia, *Lavinia Maggi, variegated*, pink and white, very large and beautiful. We hope to buy a plant in Paris or London. The roses were very good, but the flowers were all suffering from the great heat. The gardener told me that the Mespilus will not bear fruit except in the open air, and that, of course, is out of the question in England, unless it would flourish, as I think it might, in Cornwall. Our plants in the hothouse, he says, will bear flowers, but no fruit. This garden is well supplied with fresh water, which is generally scarce in the city.

We drove back to the hotel, passing through the great gate, and taking another look at the massive fortifications which surround the town. One strong wall has been raised beyond another, as the buildings spread, each a memento of a different century, marking the age of the old city, like the circles in the wood of some aged tree. It hardly realises its name to a stranger driving through the queer, steep, narrow streets. We did not see it from the water, but it must be on its *sea* side that Genova la Superba reigns supreme, and our distant view of it from the Corniche was wonderfully beautiful, with the great lighthouse flung out into the waves as a challenge or a welcome.

I forgot to tell you that we passed the entrance to the

Porto Franco, on the east side of the harbour, in our tour of investigation round the fish market, and were about to look in at the gate, when a guard motioned C. and me to retire. The ancient regulation is still in force, forbidding the entrance of priest, woman, or soldier. A strange relic of mediæval feeling, when the merchant guilds of a city were bonded together to resist the lawless soldiery or free lances dwelling in their midst in the pay of the nobles; the women and priests exciting suspicion by their tendency to intrigue and duplicity, and also, no doubt, by their love of a little private smuggling, to which their longer robes lent great facilities. At any rate, they were warned off by the old traders, whose descendants treated us in the same summary manner.

We had a long afternoon at the hotel; glad to rest and keep quiet, as the heat was excessive. At dinner there were very few people; amongst them a pleasant new arrival, a Mr. R. with his young son and daughter, from Rome, where they say the crowds of English are something wonderful. We had the 'Times' of Wednesday this afternoon (Saturday). Five ladies travelling alone and all rather near-sighted, were the only other people at the table. A gentleman whom we met at Nice told us he was trying to engage a carriage for a party of six, also all ladies! It is astonishing how many of one's countrywomen one meets all over Europe. Even the Turk who expressed his astonishment at Miss Beaufort and her sister, on the Nile, as 'a hareem belonging to nobody,' must be convinced that the English have their own manners and customs. The old joke about 'unprotected females' was worn threadbare long ago. English women have shown that they can care for themselves, and many a lonely woman, who might have been condemned by the prejudices of society to spend half her life at dull watering places, or in the gossiping circle of a little town, can now enjoy all the charm and interest of journeyings, and enrich her whole life with an experience of what is greatest and best in nature and art, sharing, as one only can by living among them, in some of the rich treasures of genius which past generations have left as a blessing to our own.

We have often noticed with what respect an English lady is always treated; her little eccentricities—if she have any—being satisfactorily explained by her nationality; and the courtesy with which these are met, proves, I think, beyond anything else, how truly the men of other nations value the pure standard and life of our women. I have heard French or German mothers commenting on the actions—sometimes foolish enough—of English girls, adding:—

'They can be trusted; they are so different, they are *English*;' that word being their unconscious testimony to the home education and influence which they see pervades, more or less, their whole lives. But I am afraid I am growing prosy, and certainly C. and I are very thankful, spite of all I have written, when we meet our countrywomen facing the difficulties of a journey alone, that we may go on our way so lovingly cared for.

This evening we had a pleasant drive of an hour, winding up a higher road and looking down on the city. The great want seems to be good boulevards or pleasant drives in the environs. The only level road leads to Sestri, and is so thickly covered with dust for most of the year, that you are almost suffocated. The immense number of carts and loaded waggons, horses, and mules constantly passing along it towards the city, crowding against each other, alone renders it unfit for ordinary carriages. The bourgeois hurry out to Sestri of an evening, to drink coffee or smoke in the little restaurants and cafés that abound there, the small hired carriages being very cheap, and you may meet hundreds on the road, crowded with bright-faced holiday makers, driving into the country, regardless of the clouds of dust. The aristocracy of Genoa, meanwhile, have only the rather dreary public gardens, round which they must drive till they are weary. We were unable to visit the noble old Doria palace, outside the Porta di San Tomaso, the gardens of which extend to the sea. The grand inscription placed by the gallant Andrea, 'Il Principe,' on the walls of his house, reads strangely in the present day, like a dead voice trumpeting forth the great deeds of the past in the midst of the busy murmur of toiling life. *Divino munere, Andreas D'Oria Ceva F. S. R. Ecclesiæ Caroli Imperatoris Catolici maximi et invictissimi Francisci primi Francorum Regis et Patriæ classis triremium IIII. præfectus, ut maximo labore jam jesso corpore honesto otio quiesceret, ædes sibi et successoribus instauravit.* MDXXVIII.'

We stopped at the door of one of the cafés to watch some people making the great camellia bouquets. They are entirely composed of *buds*, probably because when the open flower has been exposed to the sun it quickly fades and turns colour. A woman was busy preparing hundreds of crimson and white blooms, *turning back* the outer petals with the fingers, and leaving the centre ones like a little cone; each bud is fastened to a wire about three feet long, and they are then laid in circles, as close together as possible, to form an enormous nosegay, from two to three feet across! The mass of colour is so great that, even when the first freshness is over, they produce a good effect at a distance, or in the softened light of a church; but, except as decorations, they are very ugly.

Yesterday afternoon we saw a crowd round a church door, and waited to watch the procession appearing from it, headed by two banners and many silver crucifixes, and numbers of black-hooded men, the two holes for the eyes giving a horrible effect to the figures. C. and I had never seen any members of this order before—*Confraternita della*

Genoa.

Genoa

Misericordia—and we were much interested. The curé of the church had died, and his coffin had been lying there, and now these men were summoned to carry it to the deadhouse. There was a long procession of priests, and numbers of lighted candles, and then more of the black ghostly figures, bearing up the pall, each, in ludicrous mockery as it seemed to us, carrying a brilliant bouquet of red and white camellias; but probably the colours had some symbolical reference to the Passion of our Lord.

These Romish ceremonials never seem to me gorgeous *enough*, if they are to affect the soul through the senses; the trappings are so shabby, and the music so utterly feeble: I do not believe any grand procession ever *creates* a feeling that is great or noble. Every one can remember witnessing some that have stirred their hearts to the depths; but only in as far as there was a great thought or principle under the outer gewgaws, which could wake a response in them. One can well believe there was a time when all men learnt what was true and holy from the *symbols* of their creed, as nations in their infancy must be taught by outward and visible signs; and I do not think it ought to make one despise the giants of old days because they were moved by this outward show—the pomp of a Church which, in those troublous times, was the nursing mother of all truth and civilisation and loyal faith—because, now that it has ceased to be needed as a shelter and safeguard to religious belief, it fails to impress the souls of men rejoicing in the truth that has made them free, and who have learnt to smile at the tinselled splendour and gorgeous trappings, and useless ceremonials, which served as pictured lessons to the childhood of Christian faith.

We spent a quiet Sunday; the great heat making us very glad of rest. In the morning we were at the Scotch church, and heard an awfu' philosophical and argumentative sermon on the doubting Thomas, from a Scotchman, who preached in a brown coat and coloured necktie, the regular minister being absent. At the English church, in the evening, we listened to a very beautiful sermon from Mr. Strettell, on almost the same text. The two services are held in opposite houses in the Via San Giuseppe. I do not think, judging from the attendance, that there can be many English residents.

We were much interested in the expected arrival of a party of eighty from Naples, said to be students from Vienna, *en vacances*. They were to arrive by the steamer on Sunday afternoon, and had written to order dinner at seven. The whole hotel was in a state of excitement; the ordinary travellers dined in a small *salle*, improvised for the occasion—the room filled with little side-tables to accommodate us all—while the waiters hurried about half distracted, lest the great room, where carpenters were busy fitting and joining tables and seats, should not be ready in time.

It was growing dusk as we returned from church, and the room was gaily illuminated and everything beautifully prepared, down to the eighty slender bottles of *vin ordinaire* ranged round the table; the dinner having been ordered at so much a-head, and beds for as many as possible. We found the poor landlady in despair, and a little sympathy quickly drew forth her story:—

'Yes, yes, the steamer is come—and here is a pretty state of things—half of them are ladies, Germans with their wives and daughters, travelling along together, and ordering dinners and beds in this way. How am I to lodge them? Those students would have slept a dozen in a room—and it is dark already, and I can't send about hunting for beds. A precious muddle they've made with their plannings and their travelling together—ordering a dinner for eighty people, and turning us all upside down; and now, one of them comes and says he is very sorry, but so many were hungry—and they have dined on board. *Sorry*, indeed—they will have to pay all the same!'

We left Madame, who is an Englishwoman, fortifying herself, under the shock, with a sense of the justice of her cause, and mounted to our rooms, encountering many forlorn groups on the stairs, every piteous gesture asking, 'Where are we to go?'

Pallid sea-sick-looking women, in every style of crumpled costume, leaned against the walls; while stout German Hausvaters vociferated and scolded and explained. We looked down through one of the upper row of windows into the great *salle à manger*, where the candles had been put out; a feeble glimmer from a lamp showed us the long, well-appointed table undisturbed, and a solitary waiter entering, who attempted to remove the things, but a glance at their useless perfection touched him too deeply; we saw him sink into a chair, and bury his face in his handkerchief! as many a great artist of the present day may have wept when his picture was returned by a hanging committee.

Between my father's room and ours was his dressing room and a second small bed room, which we did not require; but he had locked the outer door of the former, considering it belonged to him. All the rooms were in a row, and there were doors opening into each. We had just fallen asleep, when we were roused by unearthly sounds in the chamber adjoining; moans and sighs, and then deathlike stillness. We did not know of anyone being there, and were not a little startled, but soon quieted down again into a deep sleep, when we were suddenly wakened again by the most furious knockings, and shouts. It was some time before we could collect our senses, and then we settled that some one was evidently attempting to get into the next room. The cries became frantic; the knockings redoubled; and, as there was no answer, we concluded the unfortunate inmate must have had a fit, and was unable to reply; and we became seriously alarmed. At last the house was roused; some people came up the passage, and my father, who was on the other side of the haunted chamber, was requested to rise and open the door; and through his dressing-room the visitor gained entrance, and was shut in also. Then for a few hours there was an uninterrupted murmur of talk; we tried to distinguish the voices, but it seemed to us as though a large portion of the *eighty* had taken possession of the little room. They talked all night, and I suppose slept by turns; but anyone who knows what it is to listen to German through a door, when you are trying to go to sleep, and to be startled back to consciousness when you were hoping to persuade yourself its monotony was becoming soothing, by a sudden silence and a 'So!' like the snap of a pistol, will sympathise with our sufferings. We were really very sorry for the unfortunates, though every now and then when we had dropped asleep, they woke us by a click of a key in the door, making us believe that, cramped and wearied of their prison, they were meditating a razzia, in the hope of finding a room in the hotel still empty.

We left early the next morning without encountering our neighbours, sincerely hoping the whole eighty would start in an opposite direction. May our paths never cross! Madame apologised that the early hour had prevented her procuring a bouquet to present to us on leaving.

A sketch at Porto Venere.

On the Riviera.

VII.

Spezzia, Hotel Croix de Malte : Monday.

WE are established in the most charming rooms, a suite prettily furnished, with a large bed room and dressing room, salon, and two good rooms opening one into the other for C. and me, on the first floor, for which they only ask ten francs a night! It looks so comfortable that we shall be in no great hurry to move.

We left Genoa at half-past seven this morning, and were the only passengers in the diligence, except one old man in the *intérieur*. My father had taken four places; the whole of the *coupé* for us, that we might have plenty of room, and a seat for himself in the *banquette*, where he had a pleasant companion in the *conducteur*, a good-looking, grave, quiet man, whose son is an officer in the French army, and from him my father heard an exciting story of the diligence from Nice having been stopped by seven brigands near Savona. The proprietors were warned in time, and the carriage was filled with four gens d'armes and two English passengers, who successfully repulsed the robbers, and captured four of them.

We have had a glorious day, perfect in every respect. The weather has been wonderful, a cloudless sky and summer colouring everywhere, without our having a sense of heat or even great warmth through any part of the day. The dust was a little troublesome at first, as it lay an inch and a half deep on the road, and many inches in places; but, sheltered behind our windows, we were delightfully comfortable, with plenty of room for cloaks and bags, and the carriage pockets well stuffed with fruit.

We were sorry to leave Genoa, and took many last looks at the old streets thronged with people, amongst whom were the very gentleman-like policeman, who wear well-cut frock coats and high hats, and look by far the most fashionable men on the *pavé*. One was taking his morning round in white kid gloves! We stopped at the Post Office for letters, and were much amused by the romantic-looking sentry, a gentleman with long curls, and a coat like a dressing-gown tied round the waist, leaning thoughtfully on a musket. We had five capital horses all the way, three wheelers and two to lead; the coachman drove well; and we thoroughly enjoyed bowling along the road, dashing round acute angles, one horse almost on its side, the bells jingling, the whip cracking, and the twenty feet making merry music as we went. Now and then the driver would plunge them into a gallop, and then suddenly pull up as we reached a Post House, and in three minutes a fresh team was in, a little badinage exchanged with the idlers, and we were off again.

The scenery was lovely all the way, and there were crowds of picturesque people on the roads. The carriage shook too much for me to draw as we went on, but I managed to secure some sketches during the few minutes of each change of horses, and at Chiavari we had half an hour's halt. We did not care to go into the hotel, as we preferred lunching from our own baskets, so we only bought some fresh oranges from an old man, who recommended them in very good English; he had been, I think, a sailor. For the next stage my father joined us inside, and we had a very merry little picnic, uncorking explosive soda-water out of the windows, and eating our cold tongue; then my father clambered up again to his high seat, enjoying the open air and better view. We were too happy to be anything but very quiet, both infinitely preferring the diligence to our old carriage, very comfortable as that was. It was pleasant to be able to watch the horses, the speed was so great, the changes and the little incidents of 'the road' so amusing, and we were so ex-

cessively comfortable that the twelve hours' drive hardly seemed three. It must have been like the enjoyment in the old stage coach days as compared to the slow posting of our grandparents. We found a very delicious pear, *Pere Pasciane*, called a winter pear, all along this part of the coast and at Genoa, that was quite new to us; it is valuable as being such a late keeper. The people were much amused at my sketching them, the boys screaming with delight. I drew some girls leaning over a wall, to the great interest of the men standing round, who were excessively diverted at the proceeding. The vines were more forward than any we had seen before, and growing in true Italian fashion, flung from tree to tree, as well as trained in gardens.

I made some careful observations on the style of architecture most in vogue, hoping to be able to assist J. with some valuable hints for his new house. We saw some 'sweet things' in vermilion, very simple, picked out with blue or yellow, with charming balconies and other devices painted on the walls, and here and there a broad band of ultramarine, finished off with a real vase or statue in plaster of Paris, which has a very neat effect. My father says that in one house we passed a cat was looking up from a balcony at a lady leaning down to it from a window above; the cat, the lady, and the window, being the work of a local artist!

As we drove down the steep zigzags to Spezzia the views were perfectly beautiful; the Carrara mountains flushing the softest violet against a light pink sky, their outlines marked out in clear snow-lines; the sea a faint blue, with white houses and towers dotting the shore; a cool evening breeze to refresh us, and the exhilarating pace at which we dashed round the corners, making altogether as charming a combination as it is possible to imagine.

Tuesday, the 25th.

We had dinner in our little salon, and I wrote to you before going to bed. This morning, as nine o'clock had been fixed for breakfast, we did not rise till late. We were called by the chambermaid, who brought us two lovely sprays of banksia roses from 'Monsieur.' My father had been awake early, and had been walking about the town and in the hotel garden for hours; he exceedingly enjoys this place; though yesterday's drive was saddened by many tender and mournful memories of the exceeding happiness of that journey many years ago.

We have seen no one yet in the hotel, as we breakfasted in our room.

Spezzia: Tuesday, April 25.

We have just returned from a charming five hours' excursion in a row boat to Porto Venere at the entrance of this beautiful bay, which we have seen to the greatest advantage, as the day has been perfectly fine, and yet very cool and pleasant, and the little sparkle and ripple of the water were more beautiful than an entirely calm sea. Our boat was a good one, with an awning, and we had two rowers; one an old sailor, who was boatman to Lord Byron and to Shelley, and talked to us about them, pointing out the house, on the other side of the gulf, in which the former had lived.

Cornelius O'Dowd's pathetic account of the changes here is very true, and drawn from the life, and we could sympathise with his feeling of disgust and annoyance, when we saw the tram-road and heaps of earth and stones immediately in front of the houses (one of which had been Mr. Lever's), and the great dredges with their monotonous drone as they laboured to deepen the bay close to the shore. They are making a port and an arsenal, and have very strongly fortified the island of Palmaria

Brown, the English consul at Genoa, formerly
entrance of the gulf, besides building some
·ts, and all these great works necessarily dis-
ores, though in many cases it is merely a
awback from their beauty. We coasted
t side of the bay, passing the Polla, a won-
ater spring that rises up in the middle of
ray speaks of it as being twenty-five feet in
but it appeared to us much more exten-
boatman told us that it will be at last
ins of pipes, which are to carry the water

we rowed along was most lovely, with the
r round us, and point after point of land
opening before us, Spezzia behind, and on
our left wooded hills with white buildings amongst and at
the foot of them, and above them the Carrara mountains
and the Apennines, snow-capped, and glistening against
the blue sky. We passed some Italian frigates, and
watched the huge dredges at work, and the steamers (all
English) plying between them and the open sea, where
they discharge their cargoes of mud and stones.

We landed at Porto Venere, and climbed by its one
narrow street up to an old ruined church, through the
crumbling windows of which you look down on the
Mediterranean, stretching away as far as the eye can
reach; and close below lies Palmaria, with its terraced
gardens and one or two solitary houses. In going down
again to our boat, we often stopped to admire the picturesque groups on the door steps, women with their distaffs,
and little Murillo-like children playing about. They
seemed a very poor population; but in spite of rags and
dirt, there was hardly a face that did not strike us by its
beauty of feature; and one young girl in particular, who
was plaiting a little child's hair, curled up at her feet in
the most graceful attitude, was really perfectly beautiful,

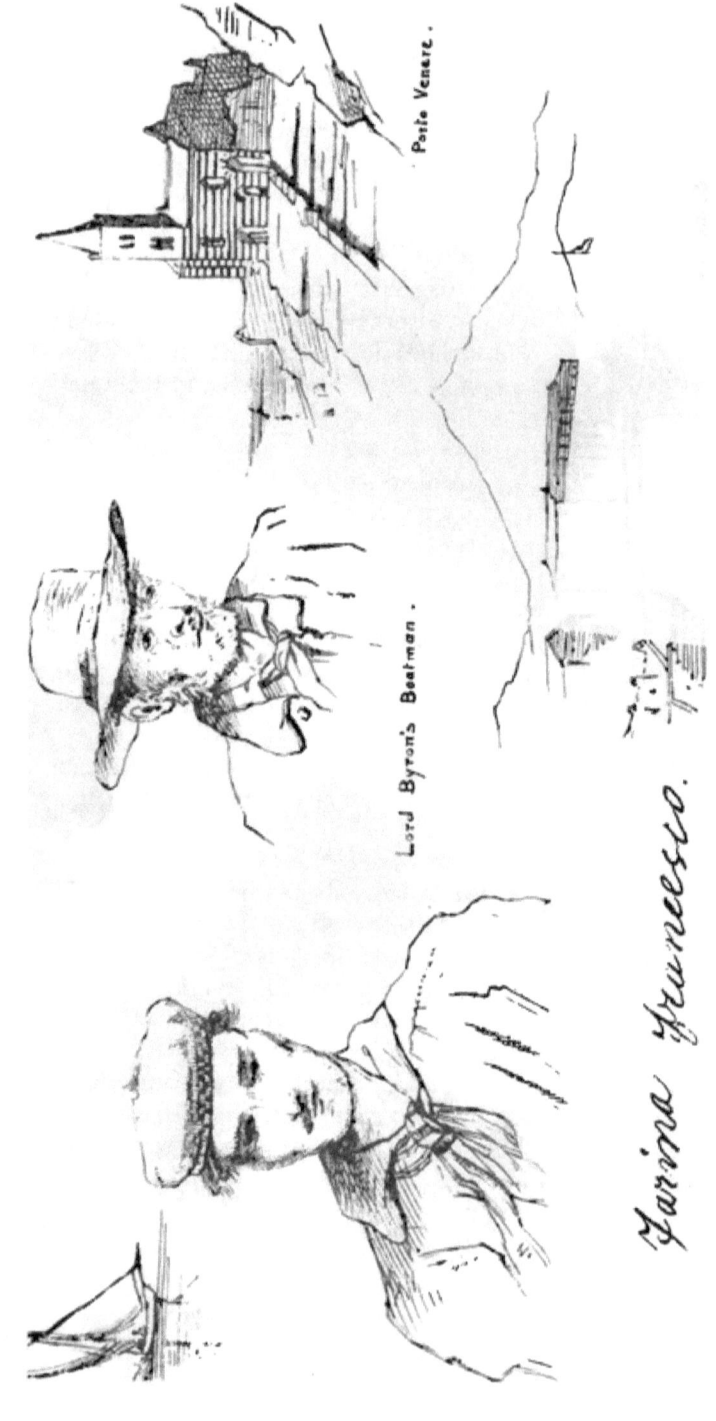

(where Mr. Brown, the English consul at Genoa, formerly lived), at the entrance of the gulf, besides building some other new forts, and all these great works necessarily disfigure the shores, though in many cases it is merely a temporary drawback from their beauty. We coasted along the right side of the bay, passing the Polla, a wonderful fresh-water spring that rises up in the middle of the sea. Murray speaks of it as being twenty-five feet in circumference, but it appeared to us much more extensive. Our old boatman told us that it will be at last utilized by means of pipes, which are to carry the water to the arsenal.

The view as we rowed along was most lovely, with the deep green water round us, and point after point of land and tiny bays opening before us, Spezzia behind, and on our left wooded hills with white buildings amongst and at the foot of them, and above them the Carrara mountains and the Apennines, snow-capped, and glistening against the blue sky. We passed some Italian frigates, and watched the huge dredges at work, and the steamers (all English) plying between them and the open sea, where they discharge their cargoes of mud and stones.

We landed at Porto Venere, and climbed by its one narrow street up to an old ruined church, through the crumbling windows of which you look down on the Mediterranean, stretching away as far as the eye can reach; and close below lies Palmaria, with its terraced gardens and one or two solitary houses. In going down again to our boat, we often stopped to admire the picturesque groups on the door steps, women with their distaffs, and little Murillo-like children playing about. They seemed a very poor population; but in spite of rags and dirt, there was hardly a face that did not strike us by its beauty of feature; and one young girl in particular, who was plaiting a little child's hair, curled up at her feet in the most graceful attitude, was really perfectly beautiful, in the features, form, and expression of her face, the finely cut mouth, the glorious eyes, and the rich Italian colouring, forming an exquisite picture, which E. tried hastily to add to her sketch book, much to the interest and amusement of the on-lookers.

On our return, we rowed close to some men who were busily collecting fish in a great net, whilst a companion, perched high up on a wooden scaffolding, gave notice, as in our pilchard or herring fisheries, of the whereabouts of the prize. Stopping at one of the small harbours we had passed before, we landed exactly underneath the large bare house which was Garibaldi's prison for two months, and saw the windows of his room, but we could not be admitted without an order from the Commandant, which it would have taken too long to obtain. We were anxious to see the *forçats*, as we had failed to do so both at Toulon and Genoa, and here we watched some of them at work, carrying water and stones, each with the heavy chain round the left ankle, and some wearing the green cap, a sign of a *life* sentence, whilst the red denotes some term of years, twenty or more. Their faces, for the most part, were not repulsive or disagreeable, and very many of them took off their caps politely, and looked quite pleasantly at us, poor things! One of the men, possibly a murderer, as he wore the green *berretta*, was sitting quietly knitting a stocking, and the sailors and people who were there talked to them and seemed quite on good terms. We asked one of the guards about them, and he told us that their work is not at all hard, principally consisting of what we saw them doing; and though they begin it at five o'clock in the morning, they are evidently not kept at it very rigorously. They have all sorts and classes here, poor men, gentlemen and priests! five hundred in all. It was a sad sight, but it interested us very much, though perhaps seeing them just as we did gave the life altogether a more favourable aspect than if we

Porto Venere

could have visited their prison itself, and watched a little more of their daily occupations. As it was, if one did not look at the chains, there seemed a good deal of liberty, though, of course, there is really the strictest watch ; and very lovely as all the natural surroundings are, one felt how weary those poor creatures must get of the very beauty, and long for any change from the ceaseless monotony of their fate ; and we rowed back to Spezzia with saddened hearts, feeling as though a cloud of sin and sorrow had dimmed the fairy-like beauty of it all.

We reached the town again about half-past three o'clock, and walked through it before returning to our hotel, but there is very little or nothing of interest in the place itself. I bought one of the curious little hats which the women wear on the forehead, or, if they are carrying any weight, on the top of the head, and which they consider a sufficient protection from the sun and heat. E. sketched our two boatmen, and asked the younger one to write his name, which he did with great pride and pleasure, though his older companion tried at first to persuade us that it would be beyond his powers !

This hotel (the Croix de Malte) is very comfortable, and so bright and clean, and pleasantly near the sea, but with a great deposit of government materials, wood, iron, stones, and earth between it and the water. Our waiter is very conversational, and tells us he was only ten when he came here, and has lived here another ten years. He recommends the food to us, and volunteered this morning to bring us 'some *miel*, which was not exactly *miel*, but *compote de pommes*'—a very accurate description, as it was a sort of apple honey, and we were glad to substitute it for the butter, which is very poor here, as in so many parts of North Italy. Milan, or rather its neighbourhood, is the great dairy for this part of the world, and along the Corniche and at Genoa the principal supply comes from there, and is of excellent quality. We see so few cows anywhere, that we wonder where they get the large quantities of milk for daily consumption, as that can hardly come from Milan also!

Close to the hotel is the public garden, which is entirely surrounded by festoons of roses, hanging from tree to tree, and very pretty in effect, in spite of the dust which has done its best to whiten them. There is a great deal of building going on, and we have been amused by watching the painters decorating the exteriors of some of the houses with the architectural and other designs so dear to the Italian heart. People can live here very cheaply and comfortably, *en pension*, and I should think most of those we have met at the table d'hôte were staying here in that way. One English lady told us that she paid ten francs a day for very good rooms, breakfast and tea served upstairs, luncheon and dinner; this, of course, is much more than in many of the Swiss or German pensions, but she seemed well content with the arrangement. At the table d'hôte, where most of the guests were Italians, was one gentleman who, we were told, had been the commander of a frigate, and before Aspromonte had placed it and his services at Garibaldi's disposal, and I suppose lost his appointment in consequence. Opposite to us was a young lady, who contrived to maintain a most animated and ceaseless conversation with her next neighbour, in spite of a lost voice, which compelled her to eke out her whispers with most wonderful gesticulations. Her dress was slight mourning, and we heard afterwards that she was a widow, her husband having died about six months ago of consumption, and in nursing him she became affected in this way, and cannot recover her voice.

We start to-morrow for Carrara and Pisa.

VIII.

Pisa: Wednesday, April 26.

ALTHOUGH I am very tired, and quite ready to go to bed, I must try to journalise a little first. We have had a most interesting day, leaving Spezzia at ten o'clock this morning in an omnibus, which took us some little distance to the railway station.

I was up early, and enjoyed a quiet walk before breakfast, bringing back with me one or two more of the picturesque little hats, which hardly vary at all in form or ornamentation, and seem even to have a regulation number of bows of red braid round the crown. We watched the people at work in the hotel garden, whilst we were waiting for the omnibus, and wondered how anything was ever grown there, unless Nature, unaided, did all the work; for the poor untidy-looking women, with their babies playing or crawling along the paths behind them, moved leisurely about, occasionally scratching a little of the surface of the ground with some impracticable-looking implement, and putting in young lettuces which must be very strong and vigorous if they ever live and flourish with circumstances so decidedly against them.

The railway runs parallel with the sea though at some little distance from it, and through a lovely valley, with blue hills on the left and beyond them a range of snow-capped mountains. The journey to the station, where we left the train in order to visit the marble quarries at Carrara, is quite a short one. The railway will soon be completed to the town itself, as well as to the port, and will be a great relief to the poor toiling oxen, who have now to draw their heavy loads some miles through the valley in addition to the steep descent from the hills.

Carriages were waiting to convey any passengers to Carrara; and we engaged one to take us there, being presently joined by a gentleman, who evidently intended to accompany us; and we found that our queer and rather dilapidated vehicle was considered an omnibus, and by no means intended for our private use; however, no one else appeared, and our companion proved pleasant, and gave us a good deal of interesting information as we wound along the dusty and rather dismal road, which brought us at length to the town, a much larger place than I had imagined. It was curious to see so many of the small houses and cottages built almost entirely of marble, and to notice that it took the place of common stone in all directions.

We drove to a very comfortable hotel; and after lunching there, started with a guide, Giulio Merli, whom my father had employed on each of his previous visits. It is possible to drive to the quarries in one of the light carts of the country; but as it would be extremely difficult in many places to pass the long string of oxen and loads of marble, and it was just the most likely time to meet them, we discarded any such idea, and tramped through the thick white dust, sometimes many inches deep, lying on the steep narrow road leading from the town, and winding up into the mountains through a really beautiful valley, with a running stream all the way. This, like everything else in its neighbourhood, is enlisted in the service of the marble, and works a succession of saw-mills, many of which we stopped to see, watching the polishing and cutting of the huge blocks, some being divided by one operation into as many as twenty-four separate layers or slabs.

Almost close to the town we began to meet the rough carts with their heavy loads of marble, and the quiet patient oxen, from two to ten or even twelve allotted to each block; but they all look fat and sleek and in good

condition; a very different state of things from that of a few years back, when they were wretched and suffering, and were cruelly used by their drivers, who sit in a very picturesque fashion backwards on the yoke between each pair, and are armed with goads, which, as far as we could judge, they do not use at all frequently or unmercifully now. The sun beat down fiercely upon us, and the white glare all around and entire absence of shade, made our climb through the soft, deep, snow-like dust very fatiguing; but our walk was constantly enlivened by the interest of watching the carts coming down, and by the conversation of our guide, who is an ardent Liberal, gave three sons to Garibaldi, and rejoices immensely over the altered and improved state of things, especially the decrease of the power of the priests and the opening of good schools. His expressions of admiration and affection for Garibaldi were almost amusing in their fervour; and he told us he considered him without an equal in the whole universe.

It was wonderful to find one's self in such a world of white marble up amongst the quarries, and strange to realise that every little stone at one's feet, and even the dust on one's boots, was of the same precious material. A very steep little path led us away from the main road, and took us, by a short descent, into a side quarry, where some men were at work with pickaxe and saw, and poor girls were carrying the loose and refuse stones on their heads to some distance, returning again and again for fresh loads, and toiling through the week at the same wearisome and monotonous labour for the miserable wages of fourpence a day! The men earn from four to five francs daily; this high rate being in part owing to the danger to which they are often exposed from the frequent accidents in blasting, and from the custom of rolling the great blocks down the sloping hill sides to the roads, when it not unfrequently happens that they crush some of the workmen in their progress. They use no cranes, and push and move the marble by means of long iron crow-bars, and even load the carts in the same way.

Some of the little Gospels were eagerly accepted, and one tall, gaunt-looking man, who was especially delighted at receiving one, exclaimed, whilst his dark, weather-beaten face lighted up with such an expression of joy and exultation, 'The priests have no power over us now; they cannot harm us any more!' Our guide begged for one for a friend of his, but we had exhausted our store then, though we were able to supply him on our return to the hotel. We carried away with us specimens of marble, the different qualities being pointed out to us; the neighbourhood of the purest and most valuable is always shown by the presence of black lines and seams in the rock.

We went back to Carrara by another and much pleasanter path, very narrow and high above the road, but green and pretty, and shaded by trees over head, whilst the stream murmured below; and then through a queer little village perched on the hill side to the regular road again, and into the town, where we visited a great many ateliers, and were much interested in seeing the *mechanical* part of sculpture, though it still remains almost equally a mystery to me how, by any amount of line and rule, they produce the result!

The studio of Signor Lazzerini was the most interesting that we visited, as we saw there a most beautiful original group, just finished, and purchased by a London dealer—'Hagar and Ishmael'—very lovely in form and expression, the utter exhaustion and lassitude of the boyish figure being wonderfully rendered. The model was made at Rome, and the sculptor is still quite a young man. This was marvellously superior to anything else he had done, unless I may except another group, which had been sent, he told us, to the Dublin exhibition. One of the sights of

Carrara is the house in which Michael Angelo was born; and close to it, in the market place, is a fine statue of his, designed for a fountain, but never quite finished. The figure is that of a river god, and is very grand.

We were very glad of a few minutes' rest at the hotel, and to have boots and clothes a little freed from their dusty covering, before starting again for the short drive to the station, where we had a long wait, and amused ourselves with 'Double Acrostics,' which we find a constant resource, whilst my father had a great deal of talk with a young engineer, who soon joined in the usual outcry against the priests, lamenting the waste of precious years spent by him, through the influence of the Jesuits, in learning Latin, when he might have devoted them to so much more useful studies; and the want of a knowledge of modern languages he especially deplored. He had two companions with him, one of whom was most hospitably intent on sharing a plate of cakes with us.

There is a flourishing little Protestant community now at Carrara, where the good seed sown in this somewhat volcanic soil is bearing fruit; the clear-minded, vigorous, but hot-headed natives, ever ready to join in any upheaving of the every-day life around them, have been won over by the living power of the Gospel; red republicans, many of them, just caught at the rebound, when, wild with joy at their deliverance from priestly bondage, it was most needful for them to be conquered by a gentler and more mighty rule.

Signor Perazzi, the minister of the Free Italian Church now established at Carrara, numbers one hundred and fifty in his congregation. The people are eager attenders of the services and the schools for children and adults; and from this little centre a wide circle of influence and civilisation is gradually spreading. An Englishman, a Mr. Robson, connected with the quarries, is a most earnest and useful helper to Perazzi, and they are both well able to assist, by their advice and experience, those who may wish, by money or personal care, to endeavour to establish similar schools and 'churches' in other small towns of Italy. I use the word in its primitive signification, as there is no building worthy of the name yet erected at Carrara; but Signor Perazzi is beginning at the right end, with a *living church* growing up around him, and there is plenty of time in the future for stone work and masonry, though the desired moment may no doubt be hastened by any Gothic devotee who would like to forward a handsome donation.

An hour's journey brought us to Pisa; the line passing through a lovely country, with numbers of little towns among the hills; and then quite suddenly and unexpectedly we came in sight of the wonderfully beautiful group of the Duomo, the Campo Santo, the Baptistry, and the Campanile. They stood apart in the wide plain, that lay brightening in the long level rays of the sun, their lengthening shadows falling sharp and clear upon the sward. The soft, fresh grass grew at their feet and hedged them round lovingly, and little trees and bushes made a sweet sense of summer greenery and peace about them, and each, bound in a companionship of centuries, seemed to bend protectingly towards the other, while wall and dome and galleried tower were illuminated by the gorgeous glow of the sunset, which turned their time-stained walls to a red gold, most beautiful against the fair blue sky. I had always fancied these buildings in the centre of Pisa, and it was so strange and startling to see them standing apparently isolated and alone in the green quiet country that lies about the city.

We had much trouble at the station with the *facchini*, who seize on the luggage and then demand exorbitant payment. Three or four of them attended us to the hotel, hanging behind the carriage where we could not see them. The landlords seem powerless to rescue their guests from the hands of these pillagers, and the police do not interfere. We were told of one case occurring here a year or two ago, in which an English physician, who refused to pay the absurd sum demanded for lifting a small portmanteau, was followed to his home by one of these men, and actually murdered in his own house! We were assured that a new system was very soon to be introduced, which would quite do away with this most disagreeable class of men, whose insolence and greediness are a real and intolerable nuisance at the end of a tiring journey.

We are at the Hotel Peverada, on the Lung' Arno, and have large rooms and a very amusing view from our windows of crowds of people and carriages passing up and down, lights twinkling all along the river and reflected in it, and on the water a little gondola with a tiny lamp, which my father is prosaic enough to call a ferry boat.

Pisa, Thursday: April 27.

We did not fulfil the idea we had entertained over night of getting up early and seeing some of the objects of interest before breakfast; but directly after it, we walked to the Campo Santo, through the quiet tranquil streets, wondering what life is really like in these places which are so utterly silent and desolate, and without stir or movement apparently until the evening, when the world swarms out like the gnats to promenade along the Lung' Arno; but Pisa is a university town, and must possess much good and intellectual society, and be really infinitely less dead-alive, spite of its looks, than many other places.

We were a long time in reaching the open space on which the four great buildings stand. I cannot tell you how curious it was really to see the Leaning Tower, not under a glass shade or on a mantel-piece, but the real original thing, quite as strange and wonderful as anything one had imagined, and slanting fully as much as the most exaggerated looking model; and when we had climbed the three hundred steps to the top, one of the great bells echoing down through the hollow centre as we mounted higher and higher, there was no doubt that we were out of the perpendicular either to the eye or foot. E. even grew giddy with the strange feeling, and did not go with us to the uppermost circle or platform, from which we had a very fine view over Pisa and the rich country round it, with the Arno winding away to the sea, and in the distance the towers of Leghorn, and far away an island which we hoped was Elba, but which turned out unfortunately to be only Gorgona. Then down the three hundred steps again, and into the Duomo; and here I stop bewildered by all the richness and beauty and interest which it is vain to attempt to describe, except most hastily, as whole days spent there would hardly suffice to satisfy eyes and mind. The church itself is exceedingly grand and rich, from the great abundance of white and coloured marble, and the gilded and elaborately decorated roof. Amongst the chief objects of interest are paintings by Andrea del Sarto; bronzes by John of Bologna; the great lamp which is traditionally supposed to have given Galileo his first idea of the theory of the pendulum, when he saw it one day set swinging; the altars designed by Michael Angelo, and on the two piers, between the nave and cupola, a beautiful St. Agnes with her Lamb, by Andrea del Sarto, and a Madonna and Child, by Pierino del Vaga. Every little corner and capital is rich in carving and beauty, but many of the older art treasures were destroyed by the great fire at the end of the sixteenth century.

The Campo Santo interested us very much, and we tried industriously to trace the subjects of some of the almost obliterated frescoes, those of Giotto being especially undistinguishable. In Orcagna's Triumph of Death, we recognised the three kings so quaintly pictured as realising its terrible power and their own poor humanity.

The whole of this strange composition is well preserved, or may possibly have been retouched or restored. Below the frescoes are old Roman and other sarcophagi, and also more modern sepulchral monuments. A bust of Count Cavour has lately been placed in the Campo Santo, with an inscription expressive of the warmest admiration and sorrow, on a tablet in the wall; and near it hang the great chains which the Pisans used to place across the river during war, as a defence against the enemy, and which, after being taken by the Genoese, and given to the Florentines, were restored by them to Pisa as a mark of a new and lasting friendship. Amongst the modern works of art are the Inconsolabile, and another still grander statue near it, a tall hooded figure, wrapped in quiet contemplation.

We visited the Baptistry last, and saw there a most beautiful pulpit, carved by Nicolo Pisano, and supported by pillars of different coloured marbles; the sides contain a series of exquisite bas-reliefs of scenes from the life of Our Lord and the Last Judgment. It is wonderfully perfect, and it is hardly possible to realise that more than six hundred years have gone by since it was wrought by those skilful hands! The floor of the building is finely and beautifully inlaid, and the great font in the centre, intended for immersion, with smaller basins at the angles, is exquisite in design and carving and work in coloured marbles. The stained glass is modern, but some French windows were very lovely. There is an extraordinary echo, which our guide awoke by a few notes in a rich, deep voice, answered apparently by a sweet-toned organ high above our heads.

We walked back to the hotel, after paying another visit to the Cathedral, and making a capital luncheon at a pastrycook's shop, for the moderate charge of twopence halfpenny, which included some most deliciously cold fresh water!

The different chambermaids are a curious study, and amuse us very much as we pass along. At Spezzia there was a cheery little person in a jacket with pockets, with piquant hair and eyes, the former in puffs, and the latter wide open. Here we are waited on by a ghost, a tall thin woman, clad in trailing garments, which are either very dirty white, or a light drab; slipshod and dusty, with a leaden-grey face and a calm unruffled manner, her lank robes hanging about her, she sweeps across the rooms—the only thing in the shape of a duster I fancy they ever see—and looks like a gigantic tea-leaf wrapped round a broom.

IX.

Lucca, Hotel de l'Univers: April 27.

WE drove to the station at Pisa at three o'clock, noticing, as we left the hotel, a letter from Garibaldi to our host, framed and ornamented, and placed in a conspicuous position on the staircase. It was an acknowledgment of the kindness and attention shown to him when passing through Pisa on his way to Spezzia, after the Aspromonte campaign.

A French paper, bought at the station, contained the sad news of the death of the Czarewitch, and in another column was the fearful and startling telegram announcing the assassination of poor Mr. Lincoln. It seems too terrible to be really true, and we are almost hoping there may be some mistake, as there are no particulars, nothing beyond the bare fact that the President is dead, and Mr. Seward not likely to recover.

It is only about an hour's railway journey from Pisa to this place, where we are comfortably established in a good hotel, and have a pretty sitting-room and an enormous bedroom for E. and me, so large that we shall hardly be able to make ourselves heard across it, with a smaller one beyond for my father. We are, I believe, the only guests, and they have therefore no table d'hôte, though we were shown the *salle à manger*, a curious, low room, round no end of corners and down dim passages and stairs, winding away, I imagine, behind other people's houses, and looking out on a square some distance from the front of the hotel.

We walked at once to the Cathedral, which is full of interest. Amongst the paintings that delighted us the most, were a Virgin and Child with Saints, by Fra Bartolommeo, the same subject by Ghirlandaio, and a beautiful picture by Bronzino, of the Presentation of the Virgin in the Temple, wonderfully rich in colouring, and life-like in the expression of some of the faces. The Altar of Liberty by John of Bologna, is also very beautiful, and interesting as an historical memorial, having been erected by the Lucchesi to celebrate the recovery of their freedom, and consecrated to Christ the Deliverer. It is in white marble, with three figures: those of Our Saviour, St. Peter, and St. Paul. From the Duomo we walked to the Church of San Romano, where a snuffy, fat old *custode* showed us a magnificent painting, by Fra Bartolommeo, called the Madonna della Misericordia, and representing the Virgin interceding for the people of Lucca during the Florentine wars.

After examining the rest of the church, we returned to the hotel for dinner; spending part of the evening in strolling on the ramparts which surround the town, and are planted with trees, making a pleasant shady walk, though not a popular one apparently, as it was quite deserted, whilst the hot dusty square, or Piazza Ducale, in front of the great Ducal palace, was crowded with citizens of all classes, listening to some occasional music from a band, stationed below the monument to the Duchess Maria Louisa of Lucca, which was erected in token of the gratitude of the city for the abundant supply of water furnished by the aqueduct, which she caused to be built. The lights and shadows amongst the trees were very pretty, but we are glad to finish our evening quietly in-doors after our busy and interesting day.

In Pisa, we met one of the hooded men belonging to a *Confraternita della Misericordia*, walking about, rattling his box in the face of passers-by, and receiving their contributions in silence. He was dressed entirely in dingy blue, and wore a large straw hat hanging on his back; such a queer, strange figure to come upon suddenly! and since we have been here, we have watched from our windows a long procession of the order, some in maize and some in unmitigated brimstone-coloured dresses, carrying a great banner through the streets, and chanting as they went.

I must leave this letter to E. to finish, so good night!

Lucca: April 28.

I cannot tell you how delighted we are with the churches; the great cathedrals are so wonderfully beautiful; and I can understand now the charm of the black and white marble, which is toned down by the rich gilding and painting, and the coloured glass; and the lines are not staring, as at Genoa, but rather two rich shades of grey, broken up, too, by exquisite tracery, and arch and columns of every size and form; the variety and richness of the wonderful mixture of architecture is quite bewildering. The outside of the buildings at Pisa and Lucca is beautiful; such deep, soft colouring, marked by centuries that have passed over the old marble! The cathedral at Genoa was striped all over, like a bournous; I am sorry if it should be vandalism to say it is unmitigatedly ugly; at all events, in our vague gropings after the beautiful, we have never reached such an ideal.

I did not know before that the 'Inconsolabile' is a real live widow, instead of a sentiment. One rather doubts the reality of a grief that can sit on a monument, like Patience, for the benefit of visitors. Storey's 'Sybil' always reminded me of this figure of Bartolini's, the same firm rounding of the limbs, one foot steadily planted in advance, the same *out-look* into the future, though with such different meaning; the Sybil, grand, incomprehensible, interpreting fate; the other, calm in her despair, because there is nothing that can terrify or comfort her any more.

It is very interesting to see the different traces of the political feeling of the people, the inscriptions everywhere on the walls, in gilt letters on public buildings, the number of the votes for the king, and over and over again, on houses, in streets and villages, 'Vittore Emmanuele nostro Re,' and 'Abbasso il Papa Re,' 'Rome for the capital,' &c.

We went to San Romano again before breakfast this morning, losing our way completely in the puzzling narrow streets, and wandering away so far that we had to apply to a sweet-faced *contadina*, who guided us back to the great square, parting from us with a stately bend of her gracious head, and a low-voiced '*Buon giorno, signorine!*'

We breakfasted, and started at ten in a good open carriage with brisk little horses, for the Bagni. It was very warm and dusty, but we thoroughly enjoyed our two to three hours' drive; most of the way the road wound by the side of the Serchio, a broad mountain stream rushing down from the hills. The land on either side is carefully protected by embankments, as, when the torrent is swelled by the melting snow, its force is very great. It was pleasant to see running water after travelling for so long through a dry country, and there were dozens of small streams falling into the large one, and a great aqueduct beside the road, carrying down water to Lucca, built by the Bourbon Duchess. The water came dashing down from the hills with such force and volume that it seemed strong enough to lift up each little bridge under which it passed, finding the openings too small for its stream, and bubbling up with explanatory splashings as it swept under each tiny arch. There were fountains in every village, and the people looked contented and flourishing, all busied in their daily avocations.

The wooded hills were beautiful with their fresh green colouring; yellow laburnums grew wild on the higher rocks, the fields were planted with barley already in the ear, shaded by chestnuts, olive, and mulberry trees, and the vines were hanging from branch to branch, the long stems and tendrils swinging in the air, and making lovely little frames for the pictures of the distant hills with their snowy tops, and the nearer ones, green and purple in the sunlight, dotted over with white villas, one of which belongs to the King, and with the great convent of the Angels, founded by the Queen of Etruria, smiling down upon the plain.

Silk worms are kept in great numbers in all the little hamlets, and mulberry trees grow in every available spot of ground. We stopped to see one very strange old bridge, the Ponte della Madalena, built by the help of the Evil One, say the peasants, who call it Ponte del Diavolo. The road across must be nearly at an angle of twenty degrees, and is paved with stones like tilted steps to keep people from slipping.

On arriving at the Bagni we stopped at Signor Pagnini's Hotel, which is much frequented by the English, who live there *en pension* at seven francs a day! While the horses were put up, we set out to visit the different little settlements. Our walk was a very pleasant one; we ascended the hill looking down on the old town, with its picturesque tower black with age, visiting the Bagni Caldi, the third village on the side of the hill, where a washed-out looking peasant woman showed us the different baths, all of marble, with that peculiar grimy appearance one always associates with such establishments. I have an intense aversion to them,—the hot steaming smell, like a great 'washing day,' the oozing slipperiness of stones and steps, the *water mark* in the great tanks, a dirty yellowish stain on the marble, salt incrustations, and a generally unpleasant sense of damp and age. Why is it that the people frequenting water establishments always have a deplorable appearance, a general want of starch and freshness? I remember encountering some in the lanes near Ilkley Wells—the gentlemen were pale and quiet, and invariably wore dress coats and suits of black, I suppose as indicative of their tone of feeling! Their effect was most mournful, as if all individuality had been packed and shampoo'd out of them.

We had a long and rather hot walk along a good level road to the more distant settlement, Bagno alla Villa, where there is an English church, and rows of good houses, *à louer*, and pretty gardens terraced up the hills. The chestnut and plane tree avenue will be delightfully shady by the end of another week. The trees were all fresh and green, the roses and purple magnolias making the gardens gay, the nightingales were singing in the bushes, and the whole place looked most inviting, and ready with its smiling welcome for all in-comers. We were literally the first arrivals, the season commencing in the beginning of May. We passed through a little street of shops and houses closed and silent, jalousies folded over the windows, doors unopened, placards and advertisements on the walls, coach-houses and livery stables, with 'pony carriages,' 'horses on hire,' 'English baker,' 'chemist,' and 'dairy,' in every variety of spelling, meeting the eye, and it was evident that the people depended largely on English custom. Indeed, during the season, there is always at least *one* resident English physician, following his patients, who flock here in crowds from Florence, while many of the Florentines themselves prefer Leghorn or Viareggio, and the Milanese go north to the valleys of the Engadine.

There were numbers of men and boys idling about watching for strangers, as a sportsman might count the moments till a flock of wild ducks comes within range of his gun, and he must have been tramping through bog and marsh land for many a weary day with an empty bag, to realise the eagerness with which these sons of the Bagni mark down their game! We walked calmly on straight ahead, and were unfeeling enough not even to require a guide, though their eyes eagerly devoured us. I think they were grateful to us only for coming to be looked at. It was like the dinner bell to a hungry man.

All the hotels and little *pensions* were scrupulously clean, the paths had been carefully swept in anticipation of the ladies' dresses, and tiny dwarf rose trees planted beside them to tempt the visitor up the steeper ones, and under every tree and at each corner were resting places for the weary or sentimental pedestrian, and roads leading up among the hills for miles around, commanding glorious views of river and valley, the chestnut woods crowning the slopes. The landlords and waiters were airing themselves, and enjoying lazily the prospect of a full and successful season. At Signor Pagnini's, whom my father greeted as an old friend, we explored the clean little rooms, with their floors of stone or well washed boards, fresh white beds and muslin curtains, everything telling of the coming hot weather, and making you grateful in anticipation. But we were thankful in reality to rest sheltered from the burning sunshine, which at midday was already sufficiently powerful.

Bagni di Lucca.

Ponte

The old Custode.

Signor Pagnini was horrified at my father's walking with us without any protection but his hat, and pressed him to take his yellow umbrella. In truth, the heat was very great and very exhausting, and I never remember feeling more footsore, the ground seeming to blister our feet, so that even a salt bath would have been endurable, if it had been one of the cold ones.

We had dinner in a shady salon, everything very well cooked and served, and then worked and talked and sketched (lazily from the windows) till five, when the carriage was ordered and we had a most pleasant drive back to Lucca. Our spirits rose with the fall of the thermometer, and we were very merry as we dashed rapidly down the winding road, passing the peasants trudging home from work, horses, loaded carts with oxen, and sometimes flocks of sheep, the most miserably attenuated little things it is possible to imagine; sometimes nearly one hundred poor little lambs, 'out at nurse,' following one thin old sheep in ignorant security to Lucca and the butchers! We watched the great rafts floated down by the stream, which here and there runs deep and strong in its narrowed channel; realising by the mass of *débris* and stones spread over a great extent of land, how frightful its ravages must be when it is suddenly swollen by the melting snow on the hills above.

The rafts, which are broken up for exportation when they reach the sea, are guided with marvellous skill by the bare-legged *bateliers*. It is exciting to watch them in a sudden current or eddy, steering with their long poles, shouting to each other, now in, now out of the water, as their unwieldy craft bob up and down, or swing half round with the sudden swirl of the stream; then in a moment they are in a rapid, and hurried and hurled along into sudden smooth water, and almost aground on a great heap of earth and stones, left in a sudden freak by the great river in one of its many paroxysms a few months ago.

And so—with many good-humoured greetings from the people in the villages as we passed, little quiet homely groupings of grandame and babies at a cottage door, mothers with their distaffs, youths and maidens talking love under the vines, weary men gathering in from field or mountain, country curés slowly wending their way to vespers, book in hand, watching the last glow of the sun on the hill tops, the chapel bells ringing their sweet evening chime—our horses, with their tinkling harness, whirled us merrily along, till the hills dwindled away behind us, and we emerged from the shadows of the higher valleys into the broadening plain, where a soft twilight was still lingering, and we saw again a low horizon, with the sun like a great crimson ball, sinking behind long black bars of cloud, the trees and hedgerows standing out dark and stiff against the glow of the western sky, and lights began to twinkle before us, and the horses' hoofs rang sharply on the pavement, as with a great snapping of our driver's whip and general flourish of trumpets, we clattered into the quiet old town, and plunging round the corner, came to a sudden full stop at the Hotel de l'Univers.

Pistoia: Saturday.

We are spending some hours here on our way to Florence to see the Robbias and some of the old churches, leaving Lucca early and breakfasting here. We walked from the station to the Hôtel de Londres; Murray describes it as 'the best; clean beds, civil and obliging people.' I am truly glad he found it so. Our edition is rather an old one, so I can only suppose the establishment has deteriorated. We mounted a very dirty stone stairs to the *salle à manger*. In second-rate Italian inns, this is always on the upper floor, the basement story being entirely devoted to kitchens and offices; in small out-of-the-way towns, the stabling department occupies the centre of the building, and the great stone hall is the general receptacle for *voitures de remise*, loads of hay, casks, logs of wood, broken crockery, fowls, and the domestics of the establishment, who accomplish the cooking in small adjacent cupboards furnished with pans of charcoal, while travellers pick their way through the crowds of loungers and masses of *débris* to the grand salon above.

We were very hungry after our little journey, but the room was so filthy, and the table so encrusted with dirt that our expectations were not highly raised; a slatternly girl brushed it over with a grimy looking cloth, which she afterwards spread on it, and served up one or two greasy looking compounds, which, spite of the old landlord's assurances that they were *molto buono*, we *could* not eat. A boy in shirt sleeves strolled in and out of the room, comfortably regardless of our presence. Having given up the breakfast as hopeless, we started to explore the town, driving about from one church to another for a couple of hours, and have been much interested in the fine old specimens of sculpture, and very much amused by the people, who, as it was market day, swarmed in the streets.

We have been more pestered by beggars here than throughout our previous journey; such persistent creatures dogging our steps, following us to the church doors, and in our tours of inspection; ragged women with piteous faces, and the same monotonous whine, '*per amor' di Dio*;' the smallest babies learning the lesson, and padding after us with their little bare feet and open dirty hands, too small to beg in words.

Fancy a man, hat in hand and screaming out an appeal to our charity, running at full speed the length of a street beside our carriage, dragging a truck loaded with straw, and what looks like a heap of rags, till you discover the figure of some poor wretch, scarcely human, strapped to the boards! These people *exploitent* their cripples, and no doubt a deformed limb represents a fixed income, like a place in the Customs. Beggary and pauperism are never considered to be a disgrace, and the church has an endless supply for the needs of her dependent children; moreover, indiscriminate almsgiving is an act of piety, and, in short, it is useless to struggle against the institutions of a country.

The steps of the Duomo, and the great Piazza itself were crowded by buyers and sellers, booths and stalls, the things exposed for sale being of the commonest description,—coarse cloths, old iron, boots and shoes, crockery, string, flour, vegetables, in great piles upon the stones. One woman was unpacking her store of bright silk handkerchiefs, paving the church steps with their brilliant colours. It is useless to describe to you any of the interiors, or the old collections of marvellous carvings and rude sculptures. The pulpits in many of these towns and cities are very beautiful. In the church of Sant' Andrea, there is one adorned with bas reliefs, illustrating scenes from the Gospels, the figures strongly marked and well preserved. It is supported on columns of red marble, resting on figures of animals, curious realistic studies, which must once, no doubt, have had some symbolic meaning, now lost to us; a lioness with her cubs, eagles and a winged lion, a lion tearing a horse, horribly true to nature. It is so impossible in hasty visits to enter into the spirit of a building, and there is so much in all of these of childish superstition and worse than childish mummery, that I think it is better, instead of trying to bring away a vague impression of the whole, to enjoy thoroughly the study of some one old work of art, tracing out the thought of the artist in each portion of his work, till your mind can travel back into his far away century, and you can look at the sculptured stories, as they grew under his hand, with a little of the same reverent love, his soul speaking to you through the quaint pious faces looking down out of the marble, none the less pathetic because here and there the corners of a ruff, or the fingers of the folded hands, have been worn away by the coarse handling of many hundred years.

There is something very touching in the passing mention of many of these old artists and their work. 'Completed by his pupil'—Murray adds the words to the brief catalogue of its merits, and a list of the stories worked into the stone. It is so easy to fill up the history; to picture the weakening hand, the failing eyesight, the earnest spirit, labouring on with patient toil, saddened at leaving his broken thought to others who may carry it out unworthily, and doubting of the fame he may gain in after years. Do you remember the lines:—

> 'And those quaint old fragments that are left us
>   Have their power in this: the carver brought
> Earnest care and reverent patience, only
>   Worthily to clothe some noble thought.
>
> But I think, when years have floated onward,
>   And the stone is grey and dim and old,
> And the hand forgotten that has carved it,
>   And the heart that dreamt it still and cold;
>
> There may come some weary soul, o'erladen
>   With perplexèd struggle in his brain;
> Or, it may be, fretted with life's turmoil,
>   Or made sore by some perpetual pain.
>
> Then, I think, those stony hands will open,
>   And the gentle lilies overflow
> With the blessing and the loving token
>   That you hid there many years ago.'

In the Duomo is the curious old tomb of the professor and poet, *Cino da Pistoïa*. He is seated in his chair, beneath a handsome Gothic canopy, lecturing to his students—stiff mediæval figures, all intent on their studies, one idler excepted, who is said to be Petrarch. At the end of the desks is the doctor's wife, *Selvaggia Vergiolesi*. The pupils are all on a smaller scale than the master, according to the pictorial idea of reverence of that age. It is an interesting fourteenth century record; the sculptured memorial of a laborious life above the sarcophagus in which the bones have rested for so many hundred years.

We drove to the ancient hospital, *Ospedale del Ceppo*, founded in 1218. The building has been modernised, and many of its ancient works of art destroyed; but there is a wonderful frieze of coloured earthenware, by the three Robbia brothers, completed early in the sixteenth century. It represents the seven works of mercy; hospitality to strangers, feeding and clothing the poor, nursing the sick, burying the dead, &c. The old monks are invariably the good Samaritans of the story, and their white and black garments are very distinguishable; sometimes a negro is introduced with a fine effect; the suppliants appear in varied attitudes, the outstretched arms often in wonderful relief, and the colouring of the whole is very brilliant. There are also circular groups beneath, the Salutation of the Virgin, and others, surrounded with festoons of fruit and flowers. Some of the Robbia Madonnas are very lovely. I like those best which are made without an attempt at colouring, there is a sweet, tender, reverent feeling about them, and the infant Christ is often child-like and natural.

We dismissed our carriage, after a little attempt at bargaining to be driven on to Florence, and thus spend the time on the road we should otherwise have to pass here while waiting for the train. But the dust half frightened us, and the carriages were not large enough for our luggage, though one hopeful little *vetturino* assured the Signore that he could take us all perfectly well; 'his little horse would do the distance (about twenty miles) in no time, and the Signore and the Signorine should have the seat to themselves, and he could sit on a board; and as to the luggage—well, the luggage was a difficulty; perhaps it was very small, and it could be hung under the seat?' His countenance fell when he heard the amount; even a Neapolitan driver could hardly take three people, and two hundred weight of boxes and bags, twenty miles in a one-horse gig after two o'clock in the day, and feel morally certain of reaching his destination by daylight. Any way, we declined the proposal, and here we are again in the dusty old inn, killing the time as best we may.

We stopped to watch an old woman cooking chestnuts over a pan of charcoal, wrapping them up, when sufficiently roasted, in a dirty, ragged, brown blanket, to keep them warm; we were too hungry to be over fastidious, and the chestnuts smelt so good, that we bought handfuls which we are now eating at our leisure, comforting ourselves with the reflection that, however disagreeable may be the cookery, we need never starve while we can have potatoes in their skins, boiled eggs, and *marrons au naturel*.

X.

Florence: Sunday, April 30.

It is almost like a dream to be writing to you from this beautiful city, of which one has heard and read and thought so much, but here we really are, in most undreamlike and substantial comfort and enjoyment, at the Hôtel de New York, on the Lung' Arno. We drove on our arrival to the Italie, but found it was full, and that rooms had been taken for us here, instead, so we are established in most excellent quarters; a charming sitting room, with a good bed room adjoining for my father, and beyond that, two very pleasant ones for E. and me, and another smaller room if we need it, and all on the first floor, up a quiet private staircase, and looking over the Arno and the Ponte alla Carraia, with the ceaseless stream of busy life passing and repassing beneath our windows.

You can fancy how we delight in the thought of being here long enough thoroughly to enjoy the rich feast of interest before us. We joined the six o'clock table d'hôte, after which we walked through the darkening streets to the Poste Restante, receiving our first impressions of the old Ponte Vecchio, the Duomo, the Uffizi (where they are making great preparations for the Dante celebration), the wonderful Palazzo Vecchio, with its beautiful Campanile, Michael Angelo's David, which not even the twilight could clothe with beauty, and the Loggia, at which we could only attempt to give a passing glance, all so full of associations with old times and people, and history, and all so curiously familiar—thanks to photographs and engravings. It is pleasant to feel that we shall soon have the more living pictures of them to carry home with us in our memories, and I am glad to have seen it all first by the dim evening light, when the details are lost in the harmony and beauty of the whole, and each tower and dome stands out so solemnly and grandly against the night sky; but even more welcome than the sight of all these were the home letters which we received through a grated window, and carried back gladly to the hotel, the lamps beginning to shine out and light us on our way.

T. W. has just called. He is detained in Florence by the illness of a friend, who has suffered, as so many others, from the Roman fever, which has been more severe this winter than usual, amounting in many cases to typhus; he himself was laid up with it at Cairo after a stay at Naples where it has been especially prevalent. We had a good deal of talk about Egypt, which seems to be in the strangest state of transition; some of the people suddenly acquiring immense wealth, others almost starving, as the price of provisions is quite fabulous, meat being sent from Marseilles and sold at the rate of a napoleon for a piece the size of your hand; cauliflowers five francs each, and the rest of course in proportion; great quantities of most expensive English machinery ordered and sent out, only to lie in barges, four deep, along the miserable quays at Alexandria, for want of a crane (which they have been *talking* of erecting for the last six years) and space to land them on, so that by and by they must be sold for old iron!

This morning we woke to heavy clouds, and soon had some refreshing rain, quite pleasant to see after a fortnight's unbroken sunshine and dust. We drove to the English church in one of the handsome open carriages which are so abundant here. The congregation was large, so many people coming up daily from Rome and Naples; and the Dante festival on the 14th also serving as an attraction.

In the afternoon, we walked to the Boboli gardens, at the back of the Pitti Palace; they are closed to the public except on Sundays and Thursdays. This has been our first disenchantment; for, instead of the orange trees and pomegranates, and the luxuriance of Southern vegetation that you expect to find in Italian pleasure-grounds, these royal gardens are of the stiffest description, with straight, steep walks, and rows of cypress trees, and sculpture in niches, with a few scattered flowers and an attempt at turf, which almost reconciles you to the damper climate and the green lawns at home; but once reach the top of these uninteresting paths, and you forget your discontent in the lovely view before you of Florence, with its background of hills, and the tower of Fiesole as a landmark amongst them, whilst the eye comes back again and again with unwearied admiration to Giotto's glorious Campanile, and the wonderful tower of the Palazzo Vecchio, an ever fresh delight in the uniqueness of its beauty. We crossed the Ponte Vecchio, and passed the Via de' Bardi, thinking of Romola. Almost all the shops are closed, and the streets are very quiet on the whole.

We have been much interested in the account of the schools here and at Naples, and people tell us it is wonderful to see how quick the children are, and how soon they emerge from their almost complete ignorance into an astonishing state of progress in reading and other things; but it is said that the last statistical report from Naples shows only one child out of every thousand of the population (in England, I believe, it is one in seven) at a school, so there is much room for improvement still.

We have two very beautiful roses on our table, like Devoniensis, only yellower. The flowers we see here for sale are exquisite; such roses and lilies, but we notice very few of any sort growing. We are enjoying being able to unpack all our things and get at our books, which make quite a little library on our sofa table.

May 1.

We have had such a lovely May day, and have just come in at eight o'clock from a delightful drive of three hours. E. and I got up in good time this morning, and walked to the flower market in the Piazza San Trinità, where we chose a glorious bouquet, making it up ourselves with bunches of lilies, pinks, orange blossom, heliotrope, lovely white roses, and a magnificent camellia in the centre, and carrying it back as a surprise for my father, and a charming adornment to our breakfast table. We devoted the morning, from ten to half-past two, to the Uffizi, and I need hardly say how great a treat it was to see this wonderful collection of treasures, many of which are so familiar, making one half fancy one must have seen the originals before. I shall carefully abstain from wearying you with expressions of admiration for pictures or statues, feeling also that one's first impressions of things may possibly alter very much, so that it is hardly wise to speak of likes or dislikes too soon. We made a general survey of the different rooms; the paintings, the Hall of Niobe, the flying Mercury, and all the other treasures, and then my father left us for a long, quiet study of the Tribune, and two or three of the adjoining rooms. Few things charm us so much at present as Andrea del Sarto's Virgin and Child, with St. Francis and St. John, and the exquisite Cardellino of Raphael, which seems to grow more and more lovely the longer you look at it; the infant Christ is so perfect, with such sweetness and child-like innocence in the face, and yet with a look of the Creator, as he holds his little hand over the bird which St. John is bringing to him.

Later in the day we drove to San Miniato, where many of the Florentines are buried, and from whence there is one of the most lovely views of the city, lying amidst its circle of hills, with the Arno winding away beyond it to the sea, and the green valley, bright with villas and gardens. Our road led us first by a long, gradual ascent through an avenue of old cypress trees, making a very pleasant shade, past Poggio Imperiale—the summer villa of the Grand Duke, now used as a seminary—through the narrowest and steepest of lanes, with hawthorn growing in the English-like hedges, and scenting the air with its white blossoms, and so up at last to the church, which is very beautiful inside, and rich in marble and mosaic work; and outside are all the tombs, so very small that we wondered how it was possible to use them, as the slabs, which fit close together and form a pavement, are only about three feet square. There are many touching little monuments, and over some children's graves were photographs enclosed in glass cases; an idea which would shock one's taste if it were not so sad!

San Miniato, according to the Florentine legend, was an Armenian prince serving in the Roman army under Decius, and suffered martyrdom, by the orders of the Emperor, under the walls. His effigies are entirely confined to Tuscany. There is an old Byzantine mosaic of the eleventh century in his church, in which he is represented in the regal crown and mantle, and holding the Greek cross. The chief objects of interest, besides these old mosaics, are the distinctive architectural features (much of the church being constructed, it is supposed, from the remains of ancient Roman buildings), some medallions by Luca della Robbia, a series of pictures representing the life of St. Benedict, and the fine windows in the recess behind the choir, each formed by a slab of Serravezza marble, which allows a certain amount of light to pass through, and must have a beautiful effect when the sun shines upon them; but as the daylight was almost past, and the twilight shadows were creeping over everything in the old church, we could not judge of this or explore the other parts as much as we should have wished.

We finished our drive by going quite round the Cascine, which is just outside the city. You enter it either by a long avenue, on one side of which is a large field, used by the English club as a cricket-ground, and beyond which is the race-course, or drive along the bank of the Arno; and here almost all the carriages congregate whilst the cooler weather lasts, and the scene is a very gay and amusing one, though the throng is much less than in the favourite 'mile' of Hyde Park, or the Bois at Paris. You soon reach the large open space before the Grand Duke's dairy, where the carriages all draw up in lines, whilst their occupants talk, flirt, listen to the music which plays twice every week, or drink the new milk, which is quite *the mode* for gentlemen, as well as ladies and children. Beyond this open space is the thicker part of the wood, and the road makes the entire circuit of it, passing under the shady avenues of great forest trees, through which the sunshine flickers and plays, whilst overhead the nightingales keep up a perfect chorus of song, every tree and bush seemingly alive with them.

But this does not offer so many attractions apparently to the Florentines as the *Piazzone*, where they can see and be seen; and, no doubt, the evening damp, the chief drawback to the charms of the Cascine, would be more felt under the heavy shade than in this open ground.

We have seen no celebration of May-day here; and we are told that very many of the usual fêtes of the Church have been altogether suspended lately, and there are wonderful changes in many ways. A large proportion of the priests wear ordinary coats and high hats, showing that they belong to the Liberal and popular party.

At twelve o'clock to-morrow, we are to be at the Bargello, where Mr. S., an enthusiastic lover of art, is going to show the recently discovered fresco of Dante to some of his friends; and we shall be glad of the chance of seeing it, as it is difficult of access just now, on account of the building being closed to the public, whilst some of the rooms are being arranged for an exhibition of mediæval art, and of objects of historical interest connected with the poet and his times.

XI.

Florence: Wednesday, May 3.

WHILE waiting for our tea, I must begin another history for you, but as I did not see C.'s last letter, you must forgive me if this should prove in some degree a twice told tale. We generally read over each other's writings to avoid repetitions, and form a wholesome check to any undue tendency to exaggeration. You can fancy what a rigorous censor of the press C. would make!

The life here is very pleasant and *very* un-English. The Florentines dine at four, and we follow their example, driving afterwards as soon as it is cool enough; the galleries and palaces are open from ten to four. We always breakfast at nine, and on two mornings C. and I tried going out for an hour before breakfast, but the heat is so great we have had to give it up, as it did not seem fair to my father to tire ourselves before starting with him, instead of being up to our day's work together.

We have just returned from our evening drive to Bellosguardo, a winding, steep road taking us to the beautiful villa and gardens. The house just now is vacant, so we were able to ascend to the roof, and enjoy the glorious view over city and plain, and distant hills, pink and soft violet with ridges of snow, and deepening into purple as the sun sank lower and their edges were cut sharp against the clearest of green and golden skies, while the long dying rays of light flushed every tower and dome of the beautiful city with their rosy glow, and everywhere masses of pink roses were flung against the dark green of the cypresses, and lemon blossom and bright coloured flowers scented the air. We sat long on the flat roof, watching the changing light, and thinking of the lines that had pictured that scene for us years ago in England.—

> 'I found a house at Florence on the hill
> Of Bellosguardo. 'Tis a tower which keeps
> A post of double-observation o'er
> That valley of Arno (holding as a hand
> The outspread city), straight toward Fiesolé
> And Mount Morello and the setting sun,
> The Vallombrosan mountains opposite,
> Which sunrise fills as full as crystal cups
> Turned red to the brim because their wine is red.
> No sun could die nor yet be born unseen
> By dwellers at my villa: morn and eve
> Were magnified before us in the pure
> Illimitable space and pause of sky,
> Intense as angels' garments blanched with God,
> Less blue than radiant. From the outer wall
> Of the garden drops the mystic floating grey
> Of olive trees (with interruptions green

> From maize and vine) until 'tis caught and torn
> Upon the abrupt black line of cypresses
> Which signs the way to Florence. Beautiful
> The city lies along the ample vale,
> Cathedral, tower and palace, piazza and street,
> The river trailing like a silver cord
> Through all, and curling loosely, both before
> And after, over the whole stretch of land
> Sown whitely up and down its opposite slopes
> With farms and villas. . . .'

I can find no other words to tell you half the charm of all that lay below us and around us, idealised in that warm evening light. As we came into the streets again and crossed the Arno, the colours of sky and river were perfectly gorgeous. It was like an enchanted land. We drove round the Cascine in the twilight, loth to return to our Hotel and tea and the other prosaics of life. The people there were anything but ideal; a gay throng of carriages in the Piazzone gathering round the Grand Ducal dairy; the gentlemen lounging about to talk and flirt and gossip with the ladies; the flower girls bringing their delicious great bouquets and sprays of blossom. How I wish you could share these lovely flowers, which are almost intoxicatingly sweet in the soft evening air, while the nightingales are singing in every bush. Fancy its being really too hot at five o'clock for us to think of driving in an open carriage, and of our actually having to wait for an hour for the sun to tone itself down!

C.'s letter was finished on Monday night, and the next morning we walked together to the Baptistry, through the quiet, shady streets, examining the beautiful doors, before the blinding sunshine should make them too dazzling for mortal eyes. The interior disappointed us after the Pisan building.

At ten we went with my father to the Uffizi, which we found closed, so we examined the grand old statues in the Loggia, watched the idlers asleep on the steps, and the people talking in the piazza, and visited the Duomo, the interior of which is not interesting, a large, bare old church, but with some gorgeously beautiful painted glass, and Giotto's and Brunelleschi's tombs, to consecrate their work; they were buried here by the Republic, the great dome rising into the blue air above their graves, the noblest of monuments to the master builder. The beautiful mosaic work of the outer walls and the rich marbles contrast strangely with the cold bareness within. Beckford truly describes it as a wonderful building that has been turned inside out.

It is very strange and almost awful to stand face to face at last with these giant art creations, which, for so many hundred years, have looked calmly down on the passions and struggles of races and peoples, gazing, with their quiet stone faces, on generations who have lived for good or evil, and passed away under their shadow, while they, strong in the beauty that God gave men power to work into their forms, are still there to make us, of this self-sufficient, self-admiring nineteenth century, feel a little humbler, each soul perhaps learning its own lesson as it traces out a mystic writing on the wall.

I could send you many words about Giotto's Campanile, only I have not one that can really make you understand its wonderful loveliness. In the great square are the statues of the two architects, still watching over their work, with Dante's marble slab, where he, living, used to sit by the stone figures of the dead, gazing with loving reverence at the great Duomo.

We joined Mr. S. and his friends at the Bargello, which is being busily fitted up just now as a sort of mediæval museum. There is one enormous vaulted chamber given up to him, filled with beautiful cabinets, carved chairs and bedsteads, gilded lamp stands, antique tables and rare china, and works in ivory and silver, and the Government has lent some magnificent pieces of Gobelins to decorate the walls. These arrived while we

In the Bargello.

Cascine. in the Piazzone

were there, and we spread them on the floor, and examined the curious old figures and brilliant colours, still very fresh and bright. One of the large chests, beautifully inlaid with coloured stones, with paintings and rich gilding, was prepared to hold the trousseau of the bride of Cosmo de' Medici, and was, I think, from designs by Giotto. It was in one of these that Rogers's hapless heroine, fair-faced Ginevra, was buried out of the daylight of love and hope. It is a pretty and sad old romance, but I always thought it was dull of her relations never to find her, and if you could see one of these beautiful great chests, you would think such a hiding impossible. They were part of every lady's dowry, and were supposed to be large enough to contain the best of her gala dresses, which could repose at full length in all the state of their stiff embroidery. Now it is too much to suppose that even the primmest of mediæval waiting women could, all through those long years, have resisted opening the box to see what store of lace or taffeta might lie concealed there.

It is altogether a man's story, and based upon false premises.

We went into an enormous room, forty or fifty feet high, which in old times was fitted up with three rows of prison cells, one above another. In the courtyard, in the centre of the building, is a well, sunk on the spot where the poor prisoners were executed; as we leaned against the pillars, looking down, we remembered Romola, and peopled the spot with the figures of those noble victims to the changing policy of parties, and the unscrupulous deeds of those old days, when men who mounted to power, climbed over the bodies of all who were brave or honest enough to care for the good of the many above the selfish ambition of the few. Though Romola herself may never have existed, such sorrows as hers still haunt the place. The walls are ornamented with rough old stone or painted shields, emblazoned with the arms of Florentines.

We came home early, and rested in our rooms, reading and drawing, as the heat was rather overpowering; dined at four, and then drove in the Cascine. Our rooms and little staircase are so conveniently near the street, that any friends we may have here can run up and knock at our door, *sans cérémonie*, and it is a great comfort not to have many stairs to climb when we come in laden with fruit or flowers. We have six windows facing the Arno, and another in C.'s room looking out on a terrace at the side, so we are well off in the prospect of an illumination.

This morning we drove at ten to the Pitti, which surpassed in interest our highest expectations. It is a strange, colossal old building, the basement formed of piles of enormous stones, rough and unequal, each block five to ten feet long and two feet in thickness, a cyclopean memorial of that proud Luca Pitti, the great Florentine rival to the Medici and the Strozzi, who began to raise this palace during the time of his short-lived power and popularity.

I cannot write to you of those glorious pictures. In one of the smaller rooms was the Madonna del Gran Duca, wonderfully lovely, but I like the Cardellino best. Raphael must have painted it after confession, it is so sweet and holy, with an inspiration of purity and love and foreshadowing sorrow; and this is only a beautiful pensive mother and a lovely child. A lady who spoke English with a foreign accent was beginning a copy. They used to be rare when the poor Duchess kept the original in her bedroom, and only allowed artists to see it when she was at her villa in the country, and often she carried it away with her in her journeys.

We came home early, looking at the jewellers' shops on the Ponte Vecchio, nestling under the dreary ancient covered passage that led from the Uffizi to the Pitti, with such queer little cupboards, built out over the water, like the houses on old London bridge, beautiful *pietra dura* work and turquoises, and strings of large and small pearls, shut into tiny glass cases which live outside the dark little room for the benefit of passengers, and are fastened into a sort of strong box at night. These pearl necklaces are bought by the Contadini, who keep them as a dowry, and they descend as an heir-loom from mother to child, and are often of exceeding value and beauty.

We have just heard that there will be a grand '*funzione*' to-morrow at San Gaetano, and some really good church music.

I forgot to tell you we went to Galileo's Museum to-day, or rather to a hall fitted up in commemoration of him and his discoveries; a fine statue, frescoes illustrating the events of his life, four busts of his chief pupils, his instruments, one of his fingers in brandy, busts of the late and ex-Grand Duke, who made the collection, were its chief interests. The adjoining Museum of Natural History is a very good one, but we did not do more than pass through some of the rooms. There is a Botanical garden attached. It is very sad to me to walk through these noble galleries and museums, and think with what delight their Ducal owners must have gazed upon their treasures, each in his turn beautifying and enriching them with added collections or single gems; and to feel how utterly they have lost it all, is not the less mournful because they were weak of purpose or narrow-souled, an effete race, that had not courage or faith or generosity enough for the age they belonged to, and have therefore ceased visibly to live where life itself is striving to become larger and freer. The servants in the royal livery sit, as they may have done year after year, watchfully at their posts, and all those silent galleries of faces look down calmly on the changes of the old palace,—fair-faced virgins; reverent angels; stern-eyed, grave apostles; Titian's beautiful women smiling from the canvas, in a golden glow; Andrea del Sarto's broad browed Madonnas, grand, solemn, awe inspiring, the great eyes half threatening, haunted with the look he read in his wife's face, and pictured so, may be, to bribe her, poor painter! or possibly hoping to work off the spell; and Doni, Raphael's friend, a quiet, earnest portrait by that great master, a strong, nervous, rather an *acid* face, that made you think the man had done well to choose that soft, tender woman, with her sweet, placid mouth and braided hair, who smiles on us again at the Ufizzi, a younger, fairer, idealised likeness, in the Holy Mother of the Cardellino; Guidos; Fra Bartolommeos; angelic Peruginos: what a wealth of beauty to leave behind!

While we were in the Pitti, I heard a lady telling Signor Pompignoli, who was copying the Seggiola this morning, that if he would be at his studio after three, ' she *calculated she'd look reound!*' The Americans at this hotel are very dismal, very silent, and hardly look like ladies and gentlemen. They are all Northerners so far; one, whom my father calls a regular fire-eater, is furious in his denunciations of the South, and, of course, of England.

Thursday.

A hotter morning than ever! We read and worked after breakfast, while my father walked to the Post Office, returning with a welcome home-letter. At a little before eleven, we were at the church of San Gaetano, which was crowded with people, but we secured good places, where we could see the altar brilliantly illuminated, and the orchestra. Two long galleries over the entrance and beneath the organ, draped with crimson, were filled by the performers, about 150 instruments and voices. The music was very good, and some of the trios wonderfully beautiful, and there was one voice of great sweetness and power. Every now and then, after a burst of the trumpets and the swell of song had died away into silence, a bell would ring, and the priests in their gorgeous dresses flitted about the altar, and some miserable little choristers uttered their feeble chant, and the people all knelt, and big candles were shaken up and down, and then suddenly a small white wand over the high desk in the gallery would be raised again, and the full tide of glorious music sweep down between the pillars, swelling and falling as though it and the incense were floated on invisible wings.

We were there altogether about two hours; there was a ceaseless crowd all the time pouring in and out, the two streams mingling where we sat in an odd collection of men gazing at the decorations, listening with absorbed attention to the music, praying on the ground, or talking to each other. Market women elbowed their way in, as though their coming there at all was in the way of business, and the few minutes spent in San Gaetano would add a centesimo to the profits on their sale, ladies with folded hands rested on their chairs, mothers with little children pressed onward, to stare and pray and listen, and jostle their way out again, lingering to gossip on the broad steps strewn with bay leaves that rustled as the people moved up and down, against the sweeping scarlet hangings at the door.

We walked from the church to the Accademia, passing the Piazza San Marco, and examined that wonderful treasury of early art. How exquisite are all the Angelicos, but I think you enjoy them quite as much in delicate engravings, or in careful copies of single figures, as in the original painting, where often the excessive use of heavy massive gilding robs the outlines of some of their purity. This is only my ignorant criticism; the colouring, of course, is most lovely, and wonderfully true and fresh in all. The very fact that these pictures are so beautiful, however rendered, proves their great and wonderful genius, whilst there is a holy reverent sweetness in the spirit of every one of them.

This Academy was established by a society of artists in the middle of the fourteenth century, the old hospital of St. Matteo being given up to it by the Grand Duke Leopold in 1784. In one large gallery, the paintings are hung in chronological order, from Cimabue and Giotto to Fra Bartolommeo.

There are some fine statues—casts, or the original models—in the entrance hall. Fedi's Dead St. Sebastian, and a beautiful Cleopatra with the Asp, are wonderful as studies of nature. We yesterday visited the collections of gems in the Palazzo Pitti; wonderful grotesques, carvings in rare stones, and gold and silver work studded with costly jewels; many fine specimens by Benvenuto Cellini, and a bronze crucifix by John of Bologna, such a marvellous work, the Christ with the head thrown forward, the lips parted, the eyes closed in death, the Perfect Sacrifice, most pathetically beautiful.

We walked back to the Hotel through the shady streets. In the evening T. W. called to show us the picture he had just bought, and we had a long talk over the brigand stories that everyone brings back from Naples. He was warned not to go to Pæstum, and applied to the Prefect, to whom he had an introduction. He strongly advised their not allowing ladies to be of the party, and confirmed all the tales they had heard. An English chemist at Naples, who has a great deal of shooting near the ruins, has been unable to go there for three years till this last season, when he could only venture by sea. The Neapolitans are constantly carried off to the mountains, and a tooth a day, or some other little souvenir, is sent to their affectionate relations as a gentle hint that a heavy bag of money will be welcome. And then these unfortunates have not the comfort of feeling, like the English, that their wrongs will be published over Europe, and that they may be sustained by unexpected sympathy and contributions. Those respectable citizens of Naples have a hard time of it just now; fancy! to lose both one's ears, or a favourite front tooth,

Unwept, unhonoured, and unsung!

T. made the excursion with Col.—— and two other gentlemen; they left their watches, and took very little money with them. Near the river, which they had to cross, there was a detachment of soldiers on guard, and they stopped to talk with the officer, asking whether there was really any truth in these stories.

'Well,' he said, 'all I can tell you is, that we shot five men in that little wood last week, and three more were hung at Salerno the day before yesterday. They are *all* brigands, and we might destroy the whole people, if we attempted to shoot every one who is a robber!'

The soldiers naturally dislike the service. People talk a great deal of the encouragement given to these miscreants by the friends of the late government, but I should doubt their having power or influence enough materially to aid, and certainly not to support, them; and probably time will show that these lawless bands are only the natural out-growth of such troublous times.

In all great changes, there must be many discontented spirits, who, hardly knowing what they really need or wish for in the future, make the new order of things an excuse for casting off old restraints and defying fresh regulations, glad enough of such a general overthrow of their world, and eager to make what they can out of the universal scrimmage. Then, too, there must be something irresistibly tempting to the wild peasants of these mountain villages, with their old robber instincts and traditions, to salve their consciences, if they have any, with a half notion that their guerilla warfare is for the good of the church and their own souls, with many pleasant fleecings of captured sheep of liberal opinions who may chance to come their way.

Poor *Italia unita*! it is an aggregate of very unruly and dissimilar atoms; and it may have, I fear, yet to pass through many fierce fires before these will be fused into the one glorious whole, the Italy of the future, in which we hope and believe.

We receive the 'Times' regularly; my father sees it at the club or the reading room, but we are very glad of it for ourselves, and for one or two of our acquaintances who are detained by illness in Florence.

After dinner, we walked quietly to Santa Maria Novella, the heat being still very great, and carefully examined many of the objects of interest in the church, the great one to us being Cimabue's picture. It was strange to look back through so many centuries to the day when that quaint old painting of the Holy Mother and Child was carried with a great rejoicing to its shadowy corner in the old church:—

> 'Bright and brave,
> That picture was accounted, mark, of old.
> A king stood bare before its sovran grace,*
> A reverent people shouted to behold
> The picture, not the king, and even the place,
> Containing such a miracle, grew bold,
> Named the Glad Borgo from that beauteous face.'

In the quiet cloisters, a grey haired sacristan was playing with a bright-faced, tiny child, a little piece of sunshine from the outer world, that had strayed in among the daisies on the grass, and the old faded frescoes crumbling away in patches under the arches round the court.

We visited the Spezieria, walking through noble rooms, rich in painting, gilding, and precious marbles, to the small 'dispensary,' where a young priest was selling bottles of scent and liqueurs and prepared waters. We bought some delicious orange flower essence to remind us in England

* Charles of Anjou.

of our great sprays of blossom, which are a daily delight here, and which, as I write, make the air of the room heavy with their perfume. Then followed an hour's drive in the Cascine, with the soft evening breeze to refresh us, and a gorgeous sunset again flushing the Arno.

As we drove back to the hotel at eight o'clock, the lights were twinkling along the banks of the river, and a golden moon shining down out of a deep blue sky, such a *night* blue as we never see in England, a rounded dome of warm purple air, with only a few soft shadows where a little vapoury cloud floated down into the red glow of the horizon. So good night again and good bye! This letter must be posted to-morrow. We are very happy here and well; you must not be critical over our journalisings; we *could* send you a page from Murray.

XII.

Florence: Saturday, May 6.

POOR Mrs. Trollope's death has been much felt here, Theodosia Trollope, who has written so much on Italy, and who was an accomplished Italian scholar, and much beloved by the Florentines. It is proposed that the municipality should place a memorial tablet on her house, as they have done at the Casa Guidi for Mrs. Browning.

We have still perfect summer weather. Yesterday morning early, we visited some of the picture dealers' studios, and saw one or two beautiful copies. Later in the day we drove to the church of San Lorenzo, which is wonderfully interesting, with the Medicean Chapel, vast and gloomy, and cold in its magnificence; a great catacomb, with the echoes of all the thousand feet that have trodden that old stone pavement haunting it, and a sense of ghostly presence that makes one shudder, when the careless *custode* lets the great door close with a sudden jar, as though the old stone Cosmo and Ferdinand might come down from their marble heights and pace a dreary round, when once the living world was locked out safely, and they were certain of twelve hours or more of silent awful darkness, till a busy sacristan, perhaps, with his broom, comes in the morning to sweep the dust off the great cenotaphs, and make all ready for wide-eyed tourists and other commonplaces of this nineteenth century.

Far more beautiful is the simpler Sagrestia Nuova, Michael Angelo's chapel, with its monuments to Giuliano and Lorenzo de' Medici. You know so well the grandeur and majestic power of these wonderful figures, that I need not try to picture them for you here. Lorenzo sitting calm and inscrutable, as if silently working out life's problems, now that death has come to end its strife. I wonder what the figures at his feet symbolise? In the new life of Florence, as the poet eyes read it from Casa Guidi windows, they had other and deeper meanings, perhaps, than the great master ever dreamt of.

> 'Michael's Night and Day
> And Dawn and Twilight wait in marble scorn,
>   Like dogs upon a dunghill, couched on clay
> From where the Medicean's stamp's outworn,
>   The final putting off of all such sway
> By all such hands, and freeing of the unborn
>   In Florence, and the great world outside Florence,
> Three hundred years his patient statues wait
>   In that small chapel of the dim St. Lawrence.
> Day's eyes are breaking bold and passionate
>   Over his shoulder, and will flash abhorrence
> On darkness, and with level looks meet fate,
>   When once loose from that marble film of theirs:
> The Night has wild dreams in her sleep; the Dawn
>   Is haggard as the sleepless; Twilight wears
> A sort of horror; as the veil withdrawn
>   'Twixt the artist's soul and works had left them heirs

> Of speechless thoughts which would not quail or fawn,
>   Of angers and contempts, of hope and love;
> For not without a meaning did he place
>   The princely Urbino on the seat above
> With everlasting shadow on his face,
>   While the slow dawns and twilights disapprove
> The ashes of his long-extinguished race,
>   Which never more shall clog the feet of men.'

The rough unfinished exteriors of most of the churches contrast painfully with the magnificence within. I think it must have been a sign of the beginning of weakness and decay in art in its truest sense, when the unity and completeness of beauty was lost, and people were satisfied with gorgeous magnificence by the side of unsightly blemishes, rough masonry built into walls stately with polished columns and delicate tracery. Of course these great buildings were the work of many generations, and it was easier for some rich churchman or mighty noble to add a little chapel, gorgeous and perfect in itself, or a tomb, to perpetuate his name, than to do his part to make the one great whole more perfect; and so these churches are an elaborate mosaic, to the honour and glory of the different workers and patrons; often without even a fitting setting, which might perhaps be added for the honour and glory of God!

We went to the Egyptian Museum to see Raphael's great fresco of the Cenacolo. It was only discovered in 1845, the rigorous order of nuns who formerly occupied the convent of San Onofrio jealously excluding visitors of any kind or degree. The painting is in his earliest manner, and many of the faces are very beautiful. The arrangement of the table and figures is similar to the paintings by Leonardo da Vinci and Andrea del Sarto, except that in this the figure of Judas is in the foreground, alone on one side of the table.

After our four o'clock dinner, and much pleasant talk with an English lady who is our nearest neighbour at table, we drove to Fiesole by the Porto Gallo, winding up by a good road between gardens and villas; the yellow Banksia roses hanging in masses on the walls, and pink rose hedges scenting the air. We passed the Dominican convent and church, which we hope to visit in another expedition; the convent, however, is closed against women. How I wished we could see it, and explore the old cloisters, beautiful with pictured Virgins, where—

> 'Angelico,
> The artist-saint, kept smiling in his cell
> The smile with which he welcomed the sweet, slow
> In-break of angels, (whitening through the dim
> That he might paint them!)'

We stopped at Mr. Spence's beautiful villa, and were disappointed at finding no one at home. However, trusting to his kindly promised welcome, we passed through the house to the cool, spacious loggia, into which the larger drawing rooms, and a music room rich with a great organ, open. The summer evenings here must have a wonderful charm to those who sit watching the sunset lights on the hills, and the domes and towers of the beautiful city far below, every little slope crowned with villas, buried in masses of green. We waited while the lovely colours changed and deepened till it grew almost too dusk to enjoy the wealth of pictures, busts, old carvings, and beautiful work in gold and bronze, in the rooms. A sleeping Cupid lying on a dog, by Fantochiotti, is a clever study of repose.

There are many interesting Medicean portraits, and Mr. Spence, delighting to revive the grand old associations of his villa, has placed a bust of Plato crowned with green bay leaves, in one of the saloons. We drove up the hill to the little quaint old town of Fiesole, and then walked up a steep paved slope, pursued by peasants with straw work for sale, ridiculous peacocks, or birds of paradise, with bead eyes and wire legs, and tails of *crêpé* straw fluttering behind, as the women hurried on beside us, screaming and vociferating in the shrillest-toned Italian. On the top we rested to enjoy the view again, now purple and misty in the twilight, realising the faithfulness of Hallam's wonderful description of this same Villa Mozzi we had just visited, which is so beautiful that I cannot resist giving it to you, as (the truth, though it sound bathos, must be written) Murray gave it to us! 'In a villa, overhanging the towers of Florence, on the steep slope of that lofty hill crowned by the mother city, the ancient Fiesole, in gardens which Tully might have envied, with Ficino, Landino, and Politian at his side, he delighted his hours of leisure with the beautiful visions of Platonic philosophy, for which the summer stillness of an Italian sky appears the most congenial accompaniment.

'Never could the sympathies of the soul with outward nature be more finely touched; never could more striking suggestions be presented to the philosopher and the statesman. Florence lay beneath them, not with all the magnificence that the later Medici have given her, but, thanks to the piety of former times, presenting almost as varied an outline to the sky. One man, the wonder of Cosmo's age, Brunelleschi, had crowned the beautiful city with the vast dome of its cathedral, a structure unthought of in Italy before, and rarely since surpassed. It seemed, amidst clustering towers of inferior churches, an emblem of the Catholic hierarchy under its supreme head; like Rome itself, imposing, unbroken, unchangeable, radiating in equal expansion to every part of the earth, and directing its convergent curves to heaven. Round this were numbered, at unequal heights, the Baptistry, with its gates worthy of Paradise; the tall and richly decorated belfry of Giotto; the church of the Carmine, with the frescoes of Masaccio; those of Santa Maria Novella, beautiful as a bride; of Santa Croce, second only in magnificence to the cathedral, and of St. Mark; the San Spirito, another great monument of the genius of Brunelleschi; the numerous convents that rose within the walls of Florence, or were scattered immediately about them. From these the eye might turn to the trophies of a republican government that was rapidly giving way before the citizen prince who now surveyed them; the Palazzo Vecchio, in which the signiory of Florence held their councils, raised by the Guelph aristocracy, the exclusive, but not tyrannous faction, that long swayed the city; or the new and unfinished palace which Brunelleschi had designed for one of the Pitti family before they fell, as others had already done in the fruitless struggle against the house of Medici, itself destined to become the abode of the victorious race, and to perpetuate, by retaining its name, the revolutions that had raised them to power.

'The prospect, from an elevation, of a great city in its silence, is one of the most impressive, as well as beautiful, we ever behold. But far more must it have brought home seriousness to the mind of one who, by the force of events, and the generous ambition of his family, and his own, was involved in the dangerous necessity of governing without the right, and, as far as might be, without the semblance, of power; one who knew the vindictive and unscrupulous hostility which, at home and abroad, he had to encounter. If thoughts like these could bring a cloud over the brow of Lorenzo, unfit for the object he sought in that retreat, he might restore its serenity by other scenes which his garden commanded. Mountains bright with various hues, and clothed with wood, bounded the horizon, and, on most sides, at no great distance; but embosomed in these were other villas and domains of his own; while the level country bore witness to his agricultural improvements, the classic diversion of a statesman's cares. The same curious spirit which led him to fill his garden at Carreggi with exotic flowers of the East—the first instance of a botanical collection in Europe—had introduced a new animal from the same regions. Herds of buffaloes, since naturalized in Italy, whose dingy hide, bent neck, curved horns, and lowering aspect, contrasted with the greyish hue and full mild eye of the Tuscan oxen, pastured in the valley, down which the yellow Arno steals silently through its long reaches to the sea.'

There are now only a few traces remaining of the ancient Etruscan city that once crowned the heights of Fiesole; we wandered down an English looking lane, to see the old stone wall, with its great fragments of cyclopean architecture, half buried in ivy and overhanging trees and flowers.

Our drive home through the warm summer air was delicious. It is nothing like the heavy heat of an August night in England; the air was fresh, but it was so *balmy* —no other word expresses the feeling—scented with evening flowers, and there was such a hush of repose over the whole earth that mere existence seemed a lazy delicious luxury. I cannot tell you which is the pleasantest of our expeditions; we certainly much preferred Bellosguardo to San Miniato; but then the sunset we saw from the former was the more gorgeous; last night's, beautiful as it was, could not compare with it.

C. and I had a little shopping yesterday on our own account, only drawing materials and 'bonbons.' We buy the best preserved fruit I have ever tasted, figs especially, for fifteen pence a pound, and these, with English biscuits, make a luncheon we prefer to any other, breakfast and a four o'clock table d'hote being not very far apart!

At Goodban's, a library and painting and picture shop, they gave me the names and addresses of some masters, and as one of these proved to be Signor Pompignoli, whose studio we had visited some days ago, I decided to apply to him, and this morning, on driving to the Pitti, found him as usual busy over a beautiful copy of the Seggiola. He is a pleasant man, with a grey beard, and a thoroughly *good* face, and speaks a little English. He will come to us, on Tuesday morning, from eight to nine. He *sends* an easel, canvas, palette, paintings, &c. to the hotel for me (lending the easel and palette) and buys brushes and paints to save me trouble, and for an hour's lesson and the long walk to and from Santa Croce, I am ashamed to say I am only to pay him three francs! At present we cannot arrange for more than three lessons. He is teaching several English ladies, one of whom leaves on Monday, when I am to have her hour. We think this kind old Signor will be quite an amusement, a pleasant link with the outer world, when some of our English acquaintances here have left.

We spent part of the morning at the Pitti, talking a little to some of the copyists. One quiet looking woman we liked to watch at work; she is painting the Madonna del Gran Duca, and must be, I think, a widow. She gave me her card—an English name—and asked me to come to her studio.

C. has a little cold and is not feeling well, but I hope it is only the great heat, which will tell upon people fresh from England, spite of every care. The sudden changes of temperature in Florence are difficult to meet safely. After days of intense heat, a fresh breeze comes down from the hills, very pleasant but a little dangerous, and after sunset we always find the Cascine damp and unhealthy with the heavy dew, and a mist under the thick trees, and exhalations from the river. The Florentines understand this climate, and adapt themselves to it. The young men whom one sees in crowds on the sunny Lung' Arno or in the Cascine, sauntering and smoking, are always provided with a light warm coat or cloak, and we amuse ourselves with watching the instantaneous spread of umbrellas, if one or two drops of summer rain splash on the pavement. My father says he is learning to accommodate his pace to the lazy lounge of the natives, who walk so slowly that they never overheat themselves, while the

Fiesole.

May-day.

The Dante festival.

The ball at the Uffizi.

English hurry along the streets, and then plunge suddenly into the dangerous coolness of old stone churches or picture galleries.

Your letters are always so welcome; we have had sixteen since we came here, a bountiful supply for one week, and yet we still want more, and can never hear too often. My father goes daily to the grating in the Piazza del Gran Duca, returning enriched by the large bundle of newspapers and notes pushed out to him. I think this great post office is the very hottest place in all Florence. I am glad to see a curtain is hung in front of the building, shading the workers within, and forming a little band of shelter for the hundreds of people who come and go, a curious study of faces, happy, eager, anxious, many lingering to read just a line before they are jostled away by the next claimant for letters, men hawking writing paper, peasants and officials, all thronging towards the iron bars.

This morning we walked to the Uffizi, and saw the workmen busy preparing wreaths of green bay leaves and paper flowers for the open-air ball. We have been driving since dinner in the Cascine, where there are many things in course of erection. There is to be another ball on the grass, and they are busy making stands for the races, and booths and shows are being put up. I fancy it will be a second-rate affair, rather interesting possibly, but anything but classical or very imposing, great crowds of people, peasants and deputations, and it will be very difficult to engage carriages or to move about at all, but if the mob is sufficiently amusing, we shall not find our own windows dull. It will probably be something like the Shakspeare fête in England.

I do not think, after all, that the king has yet arrived, but there is the strangest ignorance about his movements, and his popularity with some classes seems rather dubious; placards have been placed on the walls, offering a reward for the king! 'Lost.' 'Who will tell us where is Victor

Emmanuel,' &c., and people speak of him with shrugs and a sort of good-tempered wonderment.

It is very warm to-night, but the clouds after the sunset look a little threatening. We read some of our letters under the thick shade of the trees in the Cascine, to an accompaniment of nightingales. The Demidoff villa is closed for repairs during this month, so we shall not be able to see it. The galleries will, I fear, be shut during the three days of the festa. It will be no disappointment to us if we do not see much of the celebration, or it should prove a failure, but we shall ever hold it in grateful memory as having formed an excuse for a second week in this lovely place. In a few days, I hope, we shall join F. and his companions at Venice, and perhaps give up the time at Munich for a day or two more in the Tyrol.

My father says everything is doubled in cost since he was here a few years ago, and we hear the same thing from every one. The arrival of the Court has no doubt affected prices, and also the introduction of the decimal coinage, a franc now only going as far as a paul did formerly. Poverty-stricken nobles have suddenly discovered themselves to be in the possession of immense *rentes*: palazzi and house property generally having wonderfully risen, to the distress of many of the poorer inhabitants; carriage hire, hotel expenses, everything is nearly doubled, and from being one of the cheapest residences in Europe, Florence is becoming one of the dearest.

When Mr. E. first came to Florence, about thirty years since, he engaged the first floor of a magnificent palace, for which he offered thirty pounds per annum; the price was considered so munificent, that the owner, in a rush of grateful feeling, volunteered to repaint the whole of the inside. These same apartments, or some precisely similar, have been eagerly secured by one of the ambassadors at a rental of 700*l*. This will give you some idea of the altered value of everything.

Sunday night.

Before going to bed, I must try to add to this, but the great heat makes us very weak and lazy; this is the most utterly 'demoralizing' day we have yet had. The sky has been a good deal clouded, and rain is threatening and I hope may soon fall, as the air is intensely enervating, and it seems too much trouble even to *think*. C. is far from well, and we have had some of our Nice prescriptions made up for her. The chemists here are honest, and have thoroughly good shops,—as great a contrast to those at Nice (where they asked any fancy sum, varying it each day for the same thing) as are the doctors' charges. The English physicians and general practitioners here charge ten francs, the Italians five, for each visit. Dr. C.'s napoleon for each ten minutes' call, during my cold at Nice, was certainly exorbitant.

My father brought us lovely bunches of yellow roses, bought at the door for a penny of an old man with whom we have formed quite an acquaintance, and who claims the right of supplying our carriage in the Cascine.

We went to the Scotch church this morning, Casa Schneider, which is exactly opposite our hotel, the river running between. The room was quite filled. There was a somewhat lengthy, but very beautiful sermon, preached by a strange clergyman; Mr. Macdougall, the officiating minister at Florence, delivering the address before administering the communion.

My father's American acquaintance, the 'Fire-eater,' sat next him at dinner, his left arm bound with crape, no doubt for poor Mr. Lincoln; a large party of Southerners are generally opposite, who will not speak, and he seems to think it unbearable even to *look* at them, as his eyes are fixed on his plate, and he eats in savage silence. It must be a little awkward for him to pass the salt to people, or help himself out of the same dish with a man he would like to fire at across the table, a sort of ruthless extermination of rebels being his creed.

This evening we drove again to the Cascine, but only took the short turn, *mezzo giro*, drawing up with the other carriages for a few moments in the Piazzone. We had not the heart to toss back all the lovely flowers showered into our carriage, but kept some pink roses and one noble Cloth of Gold. The former were absolutely scented with attar of roses, having no scent of their own! We returned through Florence to the Porta Pinti, to visit the Protestant Cemetery. There were not many graves of very general interest, and we did not see Mrs. Browning's, but we recognized poor Arthur Clough's. We were touched by finding on a stone monument, 'erected by his widow,' the name D. T., sculptor, who died three years ago, aged thirty-two or three. He was an American, and doubtless she is also,—the pale quiet woman painting for her living, with whom we had been talking at the Pitti.

What tragedies fill up the bare outlines one traces out in passing through the world! She spoke so wearily of her work, and was so evidently painting to live, without pleasure in it. I wonder if she has little children to work for, who will comfort her! I hope we shall go to her studio. It made us so sad to see that the tomb had cracked, though the stone was very large and the designs good. How many pictures she may have toiled through, wretchedly paid for as they are, to erect it. We passed many stately old palaces and grand modern houses on our way back to the hotel; the streets were very quiet, the city wonderfully well ordered, and with an English Sunday look, the shops and markets closed, well dressed people driving about, little family parties walking to the Cascine, sitting in the fields there, and going home quietly and early. The streets were thronged between eight and nine, but now they are growing empty and still. Your next letters must be sent to Venice. If you have not burnt our old ones already, will you keep them for us, as we may care to read them over at some lazy moment when we are at home together in the summer.

XIII.

Florence: May 8.

OUR last letter was posted when my father and I walked this morning to the Piazza del Gran Duca, spending some time in the Uffizi again. I like to sit quietly in the Tribune, and watch the artists at work, and drink in all its marvellous richness and beauty. The statues seem to me to be too crowded, and the gorgeous colouring of the pictures lessens their power over your senses. They struck me as strangely small, hardly life-sized. If the room was quite empty the effect might be different, but an artist's easel is almost resting against the shoulder of the Medicean Venus, while tourists are seated by the pedestal, deep in the study of Murray, with a veil or feathers brushing one of her delicate feet. You long to see her raised above the profane crowd into a quieter-toned atmosphere, where her beauty might make 'a sweet immortal wonderment' for human gazers. The Arrotino and the Dancing Fawn and the Wrestlers are wonderful in their force and power and strong vitality, but I would gladly move them, too, away from the busy, curious, jostling, amusing throng of visitors, whose comments are a study in themselves, and an endless diversion, so queer and original and ignorant, as ours, too often, may sound to you.

We visited one little side room, where there are some

very beautiful Robbias in white marble, and an unfinished bas relief by Michael Angelo, and wandered down the long covered passage, wondering whether we could make our way out into the streets again, but we found a locked door at the corner by the Ponte Vecchio, and had to retrace our steps. It was very ghost-like and weird in that silent old gallery, with its windows looking on the river, and the long row of dead Medici, with the wicked old Catherine's white face and hands gleaming out amongst them, looking down on us from the walls. We had a little talk with one or two of our acquaintances, and then left, obtaining from the end window a good view of the long wreaths of evergreens beneath, which roof in the whole length of the Uffizi, or rather the space or court below.

I drove to the Via Scala, to call on Madame de Sanctis, whom I found at home. She lives in the *primo piano* of a large house. The great staircase is common to all, and there is a row of bells at the doors of most of these Italian dwellings. In some there is a porter's lodge attached to the hall, and the servant directs you, but in most cases you are expected to know by instinct on which floor the people you are anxious to see, reside. Woe to you if you are uncertain, for Italian stairs are steep and the houses high, and you have to toil up them and ring at the first *piano* on the chance, and then, after long waiting, an old woman, her head covered with a coloured handkerchief, opens it, and asks your business, and tells you, with a cross voice, her mistress is the Signora A. and that she never heard of your Signora B.; there is no such person there, no one, and the door closes in your face. Then you begin to think it may be the next house, and you climb up fresh stairs and appeal to other domestics, fruitlessly, and then another look at the card in your hand, where the address is given, sends you back in despair to the house you were in at first, and you mount again to the '*secondo*' and ring and ring till a servant comes to assure you that you are right at last. 'This is truly the Signora B.'s, but unfortunately, Signorine, la Signora is out,' and you have the satisfaction of leaving a card, and feeling you have accomplished a social duty, with considerable loss of time and patience, and a great amount of fatigue.

Even with your own servant, 'morning calls' would be laborious work in Florence, unless your capabilities for mounting stairs were large and your patience infinite, and discovering this for themselves, and feeling life would be valueless at such a price, the Florentines have wisely abandoned the custom. They are never formally 'at home.' Ladies receive calls in the Cascine, paying them to each other from their carriages, drawn up in close phalanx in the Piazzone, or chatting with the half-dozen young men who linger at their side, talking the pretty nothings of the hour.

> 'With fellow murmurs from their feet and fans,
> And *issimo* and *ino* and sweet poise
> Of vowels in their pleasant scandalous talk;
> Returning from the Grand Duke's dairy farm
> Before the trees grow dangerous at eight,
> (For, 'trust no tree by moonlight,' Tuscans say)
> To eat their ice at Doney's tenderly.'—

And to drink beer, too, later in the evening! Calls are paid again at the opera, each lady 'receiving' in her box instead of in her drawing-room; and when the music is over, the beautiful Principesse and Marchese, women of high rank and fair fame, drive to Bomboni's café, and sit in a saloon there, at one or two o'clock in the morning, drinking German and English beer with their *amanti*, their *cavalieri serventi*, who are romantic, or jocose, or witty, according to the humour of the hour, over their glasses. Their high-mightinesses, the Principi and the Marchesi, are in a room adjoining, also drinking beer, and happy that their wives should be so well amused, and in such good care! Alas for poor Italy, when these are the mothers who are to rear and guide the great men of its future! They are as ignorant and careless as children, without a child's innocence and faith.

And so I come round again to Madame de Sanctis, and my call in the Via Scala, after this long digression, but it is the new generation that she and her co-workers are anxious to reach, and raise morally and intellectually. Thus far, the middle and upper classes are unattainable, though one Florentine noble has ventured to send his daughters to the new school for ladies, established by the Moravians, where the sisters endeavour to carry out a good and liberal education; but Madame de S. has opened many schools at Turin and Genoa, by the help of her English friends, and these are going on most satisfactorily, and will, she hopes, furnish many good teachers, who may be dispersed over Italy. Since Dr. de Sanctis moved to Florence, she is unable to give them the same constant personal supervision, but she visits them from time to time, and all are still under her direction.*

She spoke discouragingly of the work of evangelization in Florence. Many over zealous workers have marred rather than aided in it, and the people are roused with difficulty. They will gladly accept Bibles and Testaments, and these are eagerly read, there being still about them a lingering aroma of the old prohibition, that makes them piquant, but the interest goes no further, and it is dangerous work rashly to shake honest belief in what we must consider error, if in return we have no power to rouse their consciences to the beauty of a purer faith and a more living gospel. I have great respect for the wisdom of those Missionaries who began with carpentering and

---

* There are more than 500 children under her care, and for the support of these schools throughout North Italy, increased funds are urgently needed. Miss Fox, Falmouth, kindly receives subscriptions or donations on Madame de S.'s behalf.

field labour, and so worked up, carrying the heathen with them, to the doctrines of a Supreme Providence and the truth and beauty of a Christian faith, instead of knocking down their idols with abstract truths and points of doctrine.

And so the workers in Italy who, I believe, in the end will meet with most success, are those who think less of proselytising than of raising and exalting the whole *moral* nature, so that, when the 'good ground' is prepared, they may learn to know the beauty of the divine.

In Florence, the power of the priests is still very great, and any real religious feeling very small. Dr. de Sanctis preached last Sunday in the Vaudois church, in Italian, to nearly seven hundred people, and they were cheered by seeing so large a gathering, but the Free Italian Church is so small and dirty, that they cannot hope to attract any one there but the very poor.

From the Via Scala I drove to the Piazza Madonna, trying to find Fabroni's, the depôt for Italian gospels, but neither I nor the coachman could make it out: then paid some more calls, one at the Hotel Couronne d'Italie, where I found Mrs. B. at home, and had a long pleasant talk with her. She may stay for the fêtes, as her child has set her heart on seeing the illuminations, and if so, we may start on the same day for Venice, and my father be able to help her at the frontier. I came home early, and C. and I had a quiet time together. She seemed better, her head well again, and chest less oppressed, and she even ventured down to the table d'hôte.

May 9.

* * Dr. W., the English physician, has been called in, and pronounces C.'s illness to be inflammation of the lungs.* * * *

XIV.

Florence: May.

* * * * * *

My daily letters will have told you of our life from hour to hour.* C.'s pleasant nurse, Carolina Bernardi, is the greatest comfort, and I go out every evening with my father for a drive, as this is essential to health, and it is pleasant to us all to feel that we can still have a little enjoyment in this beautiful city.

We have visited San Marco, and studied the Fra Angelicos in the cloister, and his wonderful Crucifixion in the chapter-house. The old saints and martyrs in the latter are so grand in their simplicity, and the great size of the whole makes it more impressive after the throng of little figures in many of his most important works. We spent some time also in the church of La Santissima Annunziata, and I was much interested in seeing the original Madonna del Sacco, which makes one dissatisfied with the Arundel Society's chromo, though, in general, I have thought their reproductions wonderfully faithful. The Madonna's face is so beautiful in the original, and so meaningless and almost ugly in the copy. The Birth of the Virgin, one of the frescoes by Andrea del Sarto, just outside the church door, is full of grace and beauty in the expression of some of the figures. The great painter is buried near his works. Here, too, is the Chapel of the Annunciation, with its miraculous picture of the Virgin, painted by Angels, and consecrated by the vows and supplications of poor human hearts, who come here yearly to lay

---

* These letters referred to my sister's serious illness, which for some weeks was an all-absorbing thought and anxiety; and, as I was unable to leave her for more than an hour or two daily, we were necessarily prevented from sharing in many of the passing interests of the time. The following etters were written as a parallel history to the journalisings of a sick room.

down their burdens of sorrow and care at her sacred feet, going away happy, and secure of a blessing; for is not the picture a holy miracle, and did not our blessed Lady's crown cost 8,000*l.* sterling, raised by very human means, and great exertion of priests and people, almost miraculous, too, in its way! Here, formerly, were stores of waxen offerings, dreadful little models of arms and legs, and babies, and horrible deformities brought by pious votaries, which were at one time so numerous that they were apt to fall at unexpected moments on the heads of worshippers, and so knocked about the pictures and silver chalices, that they were voted a nuisance by the priests, and swept away *en masse.*

I will not trouble you with details of this grand old San Marco, except just to say how lovely were Perugino's Ascension of the Madonna, and Giotto's Cross. The cloisters are wonderfully interesting, with their old memories of Savonarola; of the day when the angry people battered at the gates, and the great reformer and his brethren were dragged away to prison and a cruel death; and earlier still, when from that quiet dwelling he sent forth messages of thunder, words of half inspiration, to rouse the Church from the depth of sloth and pollution and crime into which it had fallen; when the power of his earnest words, the example of a pure life, drew many, wise and great and gifted in their different ways, to gather round him. It was this marvellous personal influence, so subtle and so irresistible, which made a Frate of a great painter, linking the name of Fra Bartolommeo with that of Savonarola.

We dismissed the carriage at the corner of the market, buying strawberries and flowers, which we carried to our hotel. The markets of Florence are poor and terribly dirty. The two principal are the Mercato Nuovo, with its famous bronze boar, and small covered space for stalls, and the Mercato Vecchio, which is held in a labyrinth of narrow streets, opening, at one end, into the Piazza Strozzi. These are so narrow that no cart can pass, and there is hardly room for two on foot between the open-air stalls. Every description of garbage and refuse is flung down upon the pavement, and putrifies in the gutter; there is no proper water supply or drainage, and no attempt at cleansing made by the people. At night, hundreds of hungry rats, large and ferocious, hold a ghastly carnival. There is no regular flower market, but the chief depôt is kept by a man in a sort of small open loggia, close to the Piazza di Santa Trinità, and he supplies most of the flower girls, who are so numerous in the city. We could find no large nursery gardens, where the flowers are grown for sale, and I believe they are chiefly supplied from the beautiful villas round Florence, their noble owners thinking it small shame to turn an honest penny.

We have been to the church of San Pancrazio. Fancy a Saint Pancras in Italy, but this poor old building was anything but melodramatic, and had not even a 'church-yard!' It is in a dark street, behind our hotel, and very near it, but we had some difficulty in finding it out at all. It has been roughly handled, and is, or has been, in disgrace, and part of the building is turned into a court of justice, into which we wandered, watching for a few moments a trial that was going on there;—an anxious prisoner guarded by gens d'armes, an officer in a high chair as witness, and a group of clever looking Italian heads busy over the law books. A sort of half-military, half-civil tribunal it looked, the gens d'armes helping to give the former air to the whole. And this reminds me that I must tell you we have discovered the especial duty of those 'exquisites' in glazed hats, whom we called policemen, here and at Genoa. The gens d'armes are the police proper, as we understand the word; the gentlemen in black are an organised body of officials, whose exact designation I do not know, but whose duty consists in languidly patrolling the city, and correcting any little matter that may be going wrong in a sanitary point of view. They attend to the general comfort, health, and well-being of their fellow citizens. For instance, if, in the course of a morning stroll, one of these should chance to remark a pot of flowers dangerously balanced on a high window sill, the law would justify him in requesting its removal, on the ground that it might prove injurious to a human head to receive its weight in passing! Though the men looked foolish and rather useless, I do not want to laugh at what may be the beginning of a gentler and better system, moral power and the weight of civil law gradually ousting the old principles of *force majeure*.

But to return to San Pancrazio, whose only interest now is, that under a part of its consecrated roof is buried away a curious and, I believe, accurate model of the Holy Sepulchre. There was very little church to be seen, and very great difficulty in finding any way into what there was, the law courts had swallowed up so much of it, but we had been directed to a cobbler in a little open shop, who was said to keep the key and act generally as *custode* and sacristan to strangers. Opposite the old flight of steps, just a stone's throw across the street, was a man busily at work, with a leather apron, and boots and shoes about the counter, and a bright-faced little wife tossing a dark-eyed baby; quite an unecclesiastical church guardian, we thought as we stopped to ask a question. 'Ah, il Santissimo Sepolcro, it was the *ciabattino*, poor man, the Signorina wanted; *her* husband was a shoemaker, *calzolaio*, but the mistake was nothing; the cobbler lived in that little stall round the corner; we were kindly welcome to the information;' and she tossed the baby just a little scornfully: class prejudices are strong in all ranks of life, and I think our mistake had a little hurt her feelings.

The little old man was sitting in a dark hole of a room,

busily at work, but started up with nervous alacrity at the first word. 'Il Sacro Sepolcro, sì, sì, Signor, it was a great sight, a very great sight,' and he hobbled quickly before us, the huge key in his shaking hand, and opened the door of an ancient chapel, which we entered with him, discerning, by the light thus admitted, a large erection of inlaid stone and marble, with fleurs de lys and other devices, in quaint black and white mosaic work on its walls. The old *custode* lit a feeble little candle, and we went in, and this was the fac simile of the Sepulchre! There was a tomb, a 'dead Christ,' life sized, a vase or urn, and one or two dim frescoes. The stone on which the figure rests was cracked and stained, and the *ciabattino* said, reverently, '*il sacro sudore!*' It is with very strange feelings that one stands in a building which ought to call up such sacred associations, but one cannot forget it is an imitation merely of a sham original, though an English bishop knelt to pray there a week or two before! How strangely different minds are affected by the same thing!

We drove away through streets beginning to be gay with flags and decorations, and crowded with people, out by the Porto Gallo, with the blue distance of beautiful Fiesole beyond, to a garden where a flower show is held for these few days of the Festa. I suppose we were too early, for there were only two visitors in the whole place: a few choice ferns, azaleas and roses, sweet smelling and pretty enough in their gay groupings, but nothing as a show collection. There was an exhibition of garden tools, and cut flowers in the usual stands, but we had soon seen all we cared to, and drove away, calling at the Accademia, to meet Mrs. B. and a very pleasant friend of hers, Signor A., a resident in Florence.

We bought Alpine and English strawberries; the latter have been introduced into Florence by an English lady, and are very fairly good and fine, but the quantity as yet offered for sale is small, and my father usually goes to the market as early as possible in the morning, to secure them. A great branch of orange blossom came with them to day, with really hundreds of flowers, sold for a penny.

As we were crossing the Piazza del Gran Duca this morning, the king passed us, returning in state from a visit to the Bargello. He was in a close carriage, so that I cannot honestly say I saw more than his finger at the window, and the backs of the three footmen who, in all the glory of their scarlet liveries, hung on behind.

There was not the least enthusiasm; not even all that blaze of red magnificence could extract a cheer from the *ragazzi* of Florence. The people looked pleased rather than otherwise, but no cry even of 'il Re' was raised, and my father was the only one who uncovered, and yet this was the king's first public appearance in his new capital! There can be no doubt that he is popular with a great number of the Italians. He is the embodiment of a great idea, and as the king of Italy the people cling to him, even while they feel that his is no kingly nature. The educated classes, liberals who value him as the head of the new system that is to regenerate Italy, speak of him with all possible respect, pass over in silence his most glaring defects and failings, and tell you of his kindness, his open-hearted ways, his desire to do his duty as king (though he does not hesitate to show how irksome those duties are to him), and find excuses for his want of polish, his rough manners, and the scandals of his life. They say, 'We know his faults, they are all out-spoken, he is honest, if he is rough, and we can trust him, he will not betray us.' Artists have told me that he is no niggard patron; though he neither knows nor cares for painting or sculpture, he will give royal commissions, as being a part of his duty. Many Italians have said to us: 'It has been a blessing to our nation to have a king who does not care to interfere with state affairs, who has wisdom enough to know when he is well served, and to leave his ministers free to act for the good of their country.' He seems centuries behind his courteous fellow countrymen, in his rough manners and mode of life, but perhaps this unpretending barbarism may be a safer example than the polished wickedness and state craft that is so often the abiding spirit of more tutored courts. The king, as every one knows, is a great sportsman, strong and brave, of unflagging energy, and with a deep love for nature in her wildest forms. His happiest hours are spent in the chase, and he may have proved a frank and pleasant companion in many a night bivouac, with men of like tastes and achievements, but it simply *bores* him to be a king, and to live in cities; and Florence, as being more of a city and further from the hunting grounds than Turin, is less pleasant to him, and he makes no secret about his dislike to the former as a residence; coming as a great *Wildschütz* who has been captured and tamed, and taught to go through his part; and thus he is carried about Florence, careless of its stately buildings and grand old memories.

In driving in the Cascine, when we have asked our coachman, 'Will the king come; is there any chance of our seeing him?' the answer would be with a shrug 'Do the Signorine care for that; per Bacco! he is not much to look at, he will not come now, he is in his beer cellar more likely!' We can only wonder how things wil be when the Court comes; at present, Victor Emmanuel is only attended by a few of his advisers, some of the ministers being still at Turin. Of course many of the old frequenters of the Pitti were adherents of the Grand Duke, many his personal friends, for he had made himself much beloved; and in society he and the Duchess were sociable and kindly; but, making large allowances for faithful attachment to a fallen master and the old *régime*, there is no reason why many of the nobles should not now rally round the king, were it not for the effect produced by his own personal character and want of royal dignity. Of course, party feeling still runs high, and from one and another one hears very different accounts of the way in which Victor Emmanuel is regarded. As the fearless, frank-hearted soldier, he must always be dear to Italian hearts; as *their* king of Italy, '*il nostro Re*,' be held in grateful honour by many for whom he has ended long years of cruel suffering and oppression. Never for one moment have Tuscans had to regret that they sought shelter under his crown. But personally he will do nothing for the social and spiritual regeneration of Italy, that strong and steady outgroping towards the light, which gives the fairest and surest promise that a new day is about to dawn for the country. We English, who remember with such thankful pride that the pure home life we value so highly has found its best exemplar in the shadow of a throne—a noble husband and wife rearing their children to love their God and their country, to seek in the tenderest ties of family life that training and those lessons that will fit them to govern others, by learning first to conquer themselves,—shrink from the type of royalty as we see it in the Italy of to-day. Politically, Victor Emmanuel has in right royal fashion been a true and gallant father to his people, the veritable *Désiré* of modern times, and it may be that his sons will carry on his good work for the Italy of the future, giving added lustre to his name by their own greatness and virtue, till the people shall mingle with their loyal devotion the knowledge, which we so thankfully cherish as a sacred memory and a living truth, that a prince may be the best, as well as the greatest, amongst his subjects.

Those poor little princesses of Sardinia have known little of the joys or shelter of a home. Their mother must have been sweet and good, and the Turinese tell stories of her gentle courtesy and kindness, as, with her little children in her hand, she would walk simply through the streets, talking to the people, or making purchases in the shops. Poor lady, it must have been hard to her to die so early, and leave all her little ones to such doubtful guardianship; perhaps she was troubled by a vision of a young bride sacrificed for her country, of a little fair child wedded to a stranger, her fate sealed between royal courts and messengers, while she was playing with her doll; of one son, growing up with vain, piteous yearnings for her love and care, though with such a saddened mist between the world and any dim light in his mind, that he hardly knew what tender ministrations he had lost; of the two bright-hearted young princes taking their places in the world, and doing their best to fill them well and nobly, with so little home shelter, with so worse than little of example of how much better and purer, and greater than other men, an ideal king should be.

The king comes into the city to transact business for an hour or two at the Pitti, and then returns to pass his day in the country. His royal sons are only state appendages, who are stationed about Italy at his pleasure; his *home* cannot be theirs. There can be no royal ladies belonging to his court, now that his daughters are married and his sons are still young; there is hardly any one closely related to him by blood but his sister-in-law, the widowed Duchess of Genoa, and she has deeply offended him by her marriage with the aide-de-camp of her late husband.

When the king entered Florence last autumn, the members of the Jockey Club hastily agreed to give him a torch-light reception. They awaited him in the street, and formed an escort to the palace; and he expressed himself much gratified by their enthusiastic greeting, the loyalty of which was all the more marked, as the weather was anything but propitious. About six weeks afterwards, on entering a room at the Club, Signor A. told us he saw a round table covered with a cloth, with a fringe of pheasants' heads, making, with a large buck lying in the middle, a very solid and acceptable present. Victor Emmanuel had shot them with his own hand, and they were sent as a graceful compliment from the royal sportsman to his friends in the Via de Legnaioli.

There was some difficulty in securing a carriage for to-morrow; the men ask exorbitant prices, and, probably, if we like to chance it, there will be plenty on hire when the time comes. However, my father thought it safer to engage one at thirty to forty francs for the day, which is considered cheap, and Mrs. B. will go with me to church in the morning. The view from our windows will be very amusing, but I dread the noise and crowds on poor dear C.'s account. Fancy the waiter coming into her room just now with great branches of iron painted with the Italian colours, forming chandeliers to be placed outside each window. The illumination seemed such a mockery when we were all so sad; I was very angry, but Francesco said it was ordered by the police, and they dared not omit it.

May 14.

As I write, the long procession is passing over the Ponte S. Trinità, with hundreds of tricoloured banners, which I can just see from our windows, people are running in the streets, thronging the bridges and windows and roofs of the houses, but there is not much shouting. The royal carriages have been dashing by, and detachments of cavalry and infantry. We feel sadly out of tune with all the noise and gaiety.

XV.

Florence: May 18.

We have not seen very much of the Festa, and yet, from one and another, I have heard many particulars. Mrs. B. joined us on Sunday morning, that we might go to church together, while my father stayed with C., and she gave us an exciting history of her proceedings. She rose at seven, and went with her maid and child to the Piazza Santa Trinità, and there, seeing an old man putting up a bench, she at once engaged it, and thus secured a good front view of the procession. You need never fear these Italian crowds, they are so gentle and courteous; there is no term of reproach that even the poorest would feel so keenly as the words, 'you have bad manners,' 'you have no politeness.' We are very much struck by the habits of the people; there is no drunkenness, no boisterous mirth. After each day of the fête, the citizens, or peasants who have come in from the country, may be seen walking quietly home, in little family groups, from the ball or the races; the women always well dressed (they pawn their things to get up well for the festa, which is all I can find to say against them); the young girls, even those who are quite poor, are always well taken care of; you never see parties of boys and girls as at English fairs; they walk with their mothers or aunts, or, when they are engaged to be married, they are allowed to go with a lover.

But to return to my story. Mrs. B. described the long procession of deputies, the banner of each town borne by its representatives. A gentleman from Venice, who stood near her, explained to her the names as they passed. The Naples flag was carried by a priest, who had been excommunicated by the Pope for daring to join in this Italian demonstration, and his course was a perpetual

triumph, the people greeting him with shouts of delight. When the Venetians drew near, bearing silently the banner of their captive city, mournfully draped and crowned with crape,—a few sad, quiet looking men, probably exiles, who had ventured to join their happier brethren, rejoicing in their freedom,—there was a burst of irrepressible feeling, and her companion covered up his face and wept.

We, of course, saw nothing of the grand scene at Santa Croce, where the king was received with immense enthusiasm, and the great Dante statue greeted with a shout of applause and admiration. We visited it one day lately, when the crowd was gone, and only a few boards lying about showed where a sort of temporary amphitheatre had walled it round. The figure is colossal, calm, and powerful, without the bitter scared look on the face, the shadows of the Inferno, and the hard lines marked by grief and exile, that so many artists have chosen to set there.

In the afternoon of the 14th we drove in the Cascine, and had a good view of the king. He is not handsome, but looks good-humoured, his face is burnt the deepest red, and he has lately dyed his hair and moustaches a jet black, instead of leaving their own honest red hue. He is short and very stout, and looks unhappy in a suit of black broadcloth. He would look better in uniform, and most at his ease in a shooting dress, when he makes a picturesque chasseur. We bought a photograph of him with his gun, which was almost handsome. In a carriage he does not appear to advantage, and he bows to the people, as though it bored him to return their greetings, as no doubt it does.

The Comptroller of his household and two other gentlemen drive with him; General la Marmora (who is lodging in this hotel), a fine, handsome man, with deep lines of thought and care on his face, always follows in the second carriage. The king's turn-out during the last few days has been quite perfect. On Wednesday he was at the races, with two elegant, well-built open carriages, coachmen and footmen in quiet livery with some dark blue about it, an outrider in scarlet, two mounted grooms in purple velvet jackets and caps following each carriage, and the horses all thoroughly good. The soldiers or mounted police clear the road where the throng of carriages is greatest; sometimes the Lancers, all fine men, with the same un-Italian light eyes and hair, and well mounted, gallop up and down amongst the crowd, making their horses rear and plunge as the people fall back.

There were some foolish circus performances going on in the Cascine, in a field close to the drive, on one of the days; a mock tournament and the trapèze, with a *loge* erected for the royal party, who, I believe, actually staid some time to witness the performances. Imagine our Queen in Hyde Park, with all London in attendance, watching a clown from Astley's walk a ball up a plank, as part of the ceremonies ordered in honour of a great poet and the unity of a kingdom!

The illuminations were wonderfully beautiful, entirely produced by oil lamps, except in front of the Hotel d'Italie, which was brilliant with gas lights. I could not leave C., but was very glad that my father should go, and he took Mrs. B. and Annie and Signor A. in the carriage for several hours, whilst I enjoyed the fairy-like scene from our windows. The lamps were hung in double rows, with slight inequalities here and there, which seemed to form arches all along both sides of the river, the long reflections of brilliant light illuminating the water. Every house was bright with little jets of flame, and above the dark roofs, far up in the still air, glistened the tower of San Spirito. Below our windows, on the right hand, I could see the windings of the Arno, as at different points along its banks the lights shone and sparkled; and the great bridge was one blaze of light, every arch marked out with a golden curve, sweeping down to join the lines of fire, which glanced up at them out of the river. Turning towards the town, there was the Ponte S. Trinità, its graceful outline burning and glowing, with a thousand bright reflections running down the stream to light up any forgotten shadows; and beyond it, and mingled in its light, the Ponte Vecchio, its houses all a blaze with lamps, and higher yet the line of brightness gilding the old Medicean passage,—a great, glad light everywhere, after the sorrow and darkness of centuries,—and far away, like a temple in the sky, over the Ponte Vecchio, the pillars and arches of San Miniato were drawn with a fiery pencil, tracing out the great platform on which it stands, where so many of the dead of Florence were sleeping in its light. Poor sad Dante! whose face haunted the festival like a ghost, how thankfully he would have accepted his exile and his wrongs, if he could have known a day would come for his beautiful city, when the Italians should gather there, as one people, to rejoice over him and Italy together.

My father returned between ten and eleven, and took me on to our bridge, Ponte alla Carraia, for five minutes, to catch sight of the Duomo and the Campanile; there was music everywhere, and the Italian colours, and a beautiful, clear night sky over it all. The people were very orderly, and before twelve the streets were quite quiet.

All the banners have been presented by the various deputies to the city, and I hope we shall see them together, as they are stored away in the Palazzo della Communità. On Monday there was not much to be seen but the races, but each day there have been lectures, readings of Dante, public meetings in the theatres, all of which we have been unable to attend. Mrs. B. kindly called, and made me drive with her for a little air in the Cascine, which was crowded with people, enjoying the shade and the coolness, and watching the horses. The king was on a sort of grand stand, but I did not see him till we met him driving away, as we did not care to go on the course; but stopped instead under the trees, enjoying the delicious coolness, and amusing ourselves with watching the faces of the crowd, while little Annie and her old French nurse ran to the railings. The horses were just starting, ridden by English jockeys; it was a wretched hurdle race, and we listened to the ominous thud of the horses' hoofs as they struck the wood in leaping. The leading man, riding a horse entered by a Neapolitan noble, was thrown very near us, and we saw the animal galloping riderless across the field. The king sent instantly to see if he were hurt, and the poor creature *walked* the length of the course, and had himself weighed out before he would give himself up to a Doctor, though his arm was broken in two places, and the shoulder blade fractured. They did not think at the time he was much hurt, but he died the next day, and we could not help feeling thankful we had had nothing to do with the races, except, of course, what we saw *en passant* in our drive. Like Longchamps, in the Bois, the 'course' borders the road. Our coachman looked rather puzzled at our odd taste in choosing a quiet, deserted alley, where we drove up and down under the dark green shade. The fresh evening air, and all that cool greenery, and the nightingales trilling out their clear, sweet songs, were such a rest to mind and body, after the long hours of anxious watching. Mrs. B. was full of the kindest sympathy; she has been such a true friend to us in our great trouble.

We drove to the dairy, and drank new milk, and filled a bottle for C. There is an old man on guard, a rough country peasant, who brings us glasses of deliciously fresh milk, and in the great stables near are long rows of sleek-looking, well-fed cows. There are two or three large buildings surrounding the Piazzone, a restaurant, and a gallery for shooting at a mark, where, at all hours of the day, one hears the report of the firing and the ping-ping of the bullets. Over the great centre stables, with their bas-reliefs of cows and oxen, is the room or rooms in which dancing took place during the popular festivals, when the Grand Duke gave balls at his dairy farm to the people of Florence.

On Tuesday evening, the Uffizi was a beautiful sight. I could not leave C. that day, but we persuaded my father to walk as far as the Loggia, to see the place lit up for the open-air ball. The rows of lamps and wreaths of flowers were very lovely, but there was no space for dancing; crowds of well-behaved people, Florentines, strangers, and contadini, filling up the place, and listening with delight to the music. On Wednesday there were races again in the afternoon, which, of course, caused an immense number of carriages to flock to the Cascine, but we kept all our front windows closed, and C. did not seem to suffer from the noise. The behaviour of the people is wonderful; we watched them returning from the racecourse; there was no drinking, no uproar or vulgarity; they looked happy and amused, and unexcited.

On Thursday evening we drove again in the Cascine, and had our two glasses of milk, as before. I am rather tired of the place, though it is pleasant enough in its quiet avenues, away from the crowd of people, who are amusing to watch for two or three times, or to mingle with, if you also have friends or acquaintances in other carriages, as often happens; but it suits us better just now to drive there than elsewhere, chiefly that we need only be away from C. for an hour daily; any of the hill expeditions requiring much more time than we could spare, and these steeper roads pass between high walls for some miles, and one has less air; while under the trees there

L

always seems to be a draught, and you meet the wind blowing from the Arno.

There is to be another great fête, early in June, to celebrate the *Statuto* instead of San Giovanni, in whose honour Florence has illuminated herself so often already. I suppose we shall be here to see it. We have resigned ourselves to an indefinitely long stay.

I have, as you will see, purposely avoided writing of C. in this journal letter, which has been scribbled at odd moments, parallel to those I have sent you daily, and when I could really write to you about nothing but the one subject that filled our hearts.

May 20.

Yesterday evening we walked to Monte Oliveto, Signor A., my father, and I. We took a carriage to the gate of the grounds, on the top of the little hill, to avoid walking through the streets; the heat is so great, even in the evening, that one soon grows weary with the least exertion. The road, like all the hilly roads round Florence, winds up between high walls crowned with roses, and through every gateway we caught glimpses of the blue hills, and the beautiful city below us. We strolled through a long cypress avenue, taking a short path across a cornfield, where we gathered a nosegay of purple iris growing wild, and looking wonderfully graceful with their bright bits of colouring amongst the long green reeds and corn, and then found ourselves close to the old monastery, the Frate resting themselves in the cool of the evening, at the door, or under the shade of the olives.

The trees were mostly destroyed, a few winters ago, during a hard frost, but new ones have been planted, and the hill retains its name. These monks claim kindred and alliance with the Eastern brotherhood of the convent of the Mount of Olives at Jerusalem. They looked quiet and inoffensive, with a sleepy, indolent enjoyment of their pleasant home, just above the din and stir of the city, and comfortable, too, in their loose white flannel dresses, and the broad shady hat. A little sloping path took us to the summit of the hill, a few yards above the building; a round piece of ground, planted with cypresses, and on every side commanding a glorious view of Florence; close to us was the bell-tower of the convent, rising above the dusky trees; sloping down far below lay a very garden of luxuriance and beauty, with houses dotted amongst the greenery, a *riant* valley circling with its broad smile the richer beauties of the old city, which was massed grandly by the side of the river, in great blocks of palaces and dwelling-houses; and farther to the right the crowded streets, with every here and there a tower or steeple shooting up towards the sky; with the domes of the Baptistery and the Duomo, the beautiful Campanile, and the grand outlines of the Palazzo Vecchio, and the Uffizi, glowing in the evening sunshine, the lengthening shadows throwing out every pinnacle and boss of stone-work, twisted column, or quaint old image, into strong relief. There had been heavy rain in the morning, and the sky still looked threatening, with banks of angry clouds behind the distant mountains, so intensely blue in the brilliant light that the ranges of hills looked grey as the olive trees against their radiant background; while on the other side, where the wide plain stretched away free of houses except the few villas that the gardens surrounded, with the Arno, like a silver ribbon, carried away towards the sea, there was clear bright light, and the soft yellow green colouring of fields and orchards. All the land is farmed by men who pay the landowners a certain percentage on the produce. The nobles, knowing little or nothing of such work in its practical details, are often terribly imposed on, but the *mezzeria*, or system of land tenure, better known as the *métayer*, prevails throughout this part of North Italy.

The air was deliciously fresh, and I thoroughly enjoyed our walk, a very short one, for the Mount of Olives is only a little hillock, the nearest of all to the city. We returned by another path, which led us into the Bellosguardo road, and so into the city again by the Porta Romana. Signor A. was going to Mr. Lever's villa for the evening (a little way beyond the gate), but he walked back with us first as far as the Casa Guidi, a dark old house, an end one, between two narrow dismal streets, but close to the Pitti. There is a little balcony, where poor Mrs. Browning used to sit and write, with the sky of Italy as the only beautiful thing her eyes could rest on. The place looked sad and desolate. Over the doorway is the white marble tablet the *Municipio* erected to her memory, a touching and graceful tribute from Florentines to the stranger who had lived among them, loving them so well: 'Here wrote, and died, E. B. Browning, who, in the heart of a woman, united the science of a sage and the spirit of a poet, and made with her verse a golden ring, binding Italy and England. Grateful Florence placed this memorial, 1861.'

She wasted away in rapid consumption, after years of delicacy, which her brilliant genius and bright earnest spirit had so gladdened, that her friends hardly realised what great cause there was for anxiety till the end was almost come. She died quite suddenly, clinging tenderly to the last to the people and city who have immortalised her name.

I like these mural tablets, when, instead of being built into cold corners of old churches, they are placed in the *homes* where men and women laboured for joy or sorrow. And this is a noble tribute to a good woman, whose memory is cherished as a wife and mother, and whose genius only consecrated home love, and duty.

We walked slowly home, stopping to eat ices at Doney's, and watch the brilliant Florentines in their carriages at the door, their servants hurrying in and out with glasses, while a crowd of gentlemen smoked and chatted in the great room or on the pavement. This morning I have had, as usual, a two hours' lesson from Signor Pompignoli, which I secure by rising at six. We breakfast at ten, but I have a cup of milk early. I have been to Mrs. R.'s room to wish her *bon voyage*; she and her invalid son are just starting for Bologna, on their way to England. We shall be sorry to lose her, after meeting daily for so long. Her maid has been very kind, and the courier, who is Swiss, told Carolina that he would come at any time to help. 'Tell your young ladies I can sit up at night, and warm the soup, and do everything, and they may send for me in a minute!' We shall anxiously watch for tidings of their journey. It is wonderful how good people are all the world over; we knew no one here when we first came, and now we have so many friends. It is almost worth realising trouble and suffering in a strange country to feel how much love and kindliness there is in everyone to make it seem to us like home. Our maid tells us, when she goes down to her breakfast, the servants of the hotel and the old porter ask with so much feeling, 'How is the Signorina? how is the poor young girl?' with an earnest 'Thank God for it,' when nurse says she has slept better.

We hear a great many stories of the old spirit-rapping furore at Florence, which was very rife a few years ago, when the American medium, Home, was taken up by the English here; he exercised a strong influence over poor Mrs. Trollope. She was a bright, vigorous old lady, a wonderful pedestrian, walking often more than five miles at a time when she was more than seventy years of age; though after sunset she would become quite blind, and had to depend on the help of her companions. Mr. Power, the sculptor, who feels strongly the baneful influence of spiritualism in America, says that the statistics of his country show that two-thirds of the inmates of asylums there have lost their reason from being mixed up with these wicked spirit manifestations! . . .

* * * * * *

* * I am afraid this letter will read to you as a kind of very disjointed patchwork, as it is only made up of a week in a sick room, and an hour's daily drive or walk. I have a bright account of Mrs. B. from Venice; we miss her very much.

May 22.

We have been out for a little while this morning, walking across our bridge, and a long way down the Via dei Serragli, to Mr. Power's studio, where we spent a most interesting hour or more. He is a noble-looking man, with finely cut features, grizzled hair under his artist's velvet *berretta*, and an eye wonderful in its brightness, and keen brilliancy of genius, softening into kindliness when he speaks. We carefully avoided politics, having been warned it was dangerous ground, and forgot he was a redhot Federal in his pleasant genial atmosphere.

We were very much struck by the eyes of his statues, differing from any we could remember in the works of other men. I cannot recollect seeing any especial criticism of what to us seemed almost a unique adaptation of nature in art. The pupil is not marked by an indented line, as in some of Gibson's works, where he cuts or draws them in the marble, but the *modelling* of the eye is so perfect, that you see the pupil there, turned to the right or left, according to the direction of the face. It is impossible to realise, without seeing it, how such an effect can be given by the simple curves of the lines of the eye itself. We saw some beautiful ideal heads, Ginevra, Diana, Psyche, but with the exception of the Greek Slave and the 'California,' which you must remember at the Great Exhibition, and the Listening Boy with a Shell, none of his works seemed to me to show great *genius*. In these beautiful and fanciful heads, it was very much the same face over and over again, with graceful treatment of the hair; but if the names of all had been changed, they would have been none the less expressive. His 'Eve hesitating,' with the apple in her hand, I liked; and there was a second, 'Eve disconsolate,' and an allegorical figure of America triumphant, but nothing one very much cared for, or that looked like a great piece of his own life and soul grown into marble.

His busts of living people are very clever; some grand old Greek and Roman-like heads of American statesmen, and one of his wife, a fine interesting face. Everyone agrees in the marvellous fidelity and truth of his portraits, and I should fancy his real power lay in his grasp of the meaning of each line and wrinkle time and life have written on the human face, finding out the soul and thoughts of the man, and working that into the marble with pre-Raphaelite fidelity and care. People say that his portraits have a weary, saddened look; he is so earnest to make his work good and perfect, that he exacts long and frequent sittings, and sometimes the expression of resigned, but hopeless, patience is transferred to the marble!

Mr. Power obtains high prices for his busts; he showed us studies of his daughters' heads, fine-featured, handsome women; the only one of his children who inherited his taste and genius was a daughter, who died not very long ago. We were delighted with a study of one of his children's hands as a baby, a little 'pud' growing out of a sunflower; and it was pleasant to hear him talk of the little model, as if he loved every tiny fat finger tip with all his heart.

We parted great friends, and he gave me a bouquet of roses from the studio garden. We shall go to see him again, and meanwhile we have bought cartes of two of the busts, and a good likeness of himself from his sons, who live close to him, and are photographers by profession. He is going to add his autograph for me, and offered to go with us at once to the sons' house to write it, but this of course we would not allow. His last work is a head of Christ, well executed, but without any striking point about it. There is the usual type of feature, and hair, and the full eastern eye, but it is wanting in expression; there is little of compassionate love or sorrow or divinity in the face.

We walked quietly home, through blinding dust and glare and heat, which becomes more oppressive every day. C. and I are re-reading 'Romola' together; the descriptions are so true, it is like an idealised Murray—an old-world guide to Florence; just as the city, in its turn, is a daily illustration to the book—grand stone-pictures in *alto relievo*, which we enjoy afresh with every page.

The table-d'hôte dinner, as usual, was at four o'clock; there is another at seven, but that is either too late or too early, when one drives of an evening. Each day we see the numbers diminishing. A furious Northerner and his meek wife, and decided and rather strong-minded little daughter, sat next us as before; whilst opposite were a German couple; middle-class, agreeable people. He, a sensible, ugly little man, speaking broken English, she a fair German madonna, who smiled at us with the eyes of an angel over her toothpick! There were other Americans at the lower end of the table, but some of those who sat near us yesterday were gone. There was a great deal of bloodthirsty talk among them then, and exultation over the news of the proclamation and the reward offered by President Johnson. They all speak with bitterness of England: we meet educated and wealthy Americans, as well as the mere backwoods specimens, but with all it is the same. I am afraid this hotel, from its name, is a favourite one with Northerners, and we are rather weary of them.

I must tell you a dreadful story of Doctor W.'s, as a specimen of the feeling. There is a poor wretched man dying here from the effects of drink, who formerly served with Garibaldi. He said to him a few days ago: 'I want to give you a ring, doctor; you've been very kind to me, and this ring's worth having, I can tell you. That's made of the bone of a rebel!' Doctor W. exclaimed in horror. 'Oh!' he said, 'that's nothing; I have a regular set of them; my father sent them me over, shirt-studs and links, all to match; you are very welcome to it.' This sounds like a fable, but I saw that very ring this morning—carved bone—a large signet-ring with his monogram in the centre. What a terrible keepsake from a dying man!

XVI.

May 24: morning.

WE drove one evening with Signor A. to the Torrigiani Gardens. It is a large and beautifully laid-out piece of ground, within the city walls, with two or three pleasant houses facing the lawns and shrubberies. The family Palazzo is in another part of the town—the residence of the Marchese. This property belonged to a younger brother, Carlo Torrigiani, who died a few months ago, leaving it to his young nephew. I think the great beauty of Italian gardens has been exaggerated; after having seen some of the most celebrated in North Italy—the villas and gardens round Florence, Genoa, and Como—I cannot recall any that will really bear comparison with the well-kept grounds of an English country-house. There is a great charm in the first sight of southern vegetation, and in seeing plants and flowers and trees blooming and flourishing in the open air in wildest luxuriance, that are generally known among us as little stunted specimens, carefully cherished under glass, but where the flowers are most beautiful, they fade quickly under the hot sun which forces them into existence, so that the gayest beds are seldom as *brilliant* as a whole, and never have the same delicious freshness, as English flowers. The great defect, to our eyes, must always be the absence of that smooth elastic turf, which is the great beauty of our gardens; and that look of careful and perfect finish about everything which you so rarely find abroad. In Cornish gardens, where the mild air is so favourable for acclimatising plants, you see almost all those that so charm you in Italy, growing in nearly equal perfection. Camellias nine feet in height, and fifteen in circumference; lemons and citrons ripening in the open air —the latter often as large as a child's head—as splendid specimens as the best of those you may buy at Nice, or along the Corniche. Do you not remember the orange tree in our grandmamma's garden, trained against the wall, on which the large ripe fruit grew as plentifully as peaches; but then it was protected with glass during the most severe months of winter. Rhododendrons covered with blossom, eighteen or twenty feet high, and measuring between one and two hundred feet round the lower branches; bamboo, camphor-plant, and American aloe with its sceptre of rare blossoms, growing in perfection, make many of our southern English pleasure-grounds as beautiful as the far-famed gardens on the sunny slopes of Italy.

The terraced Italian gardens owe a great deal of their beauty to the lovely views they command, and to the *form* of the ground. One is constantly surprised by fresh walks and glades of flowering shrubs; little temples and grottos, shining out under a wealth of luxuriant creepers, where everything had seemed to be at an end; and the old statues and belvederes, which are often so grotesque and out of keeping with smooth lawns and English oaks and elms, look at you from the shadows of the cypresses in perfect accord with the spirit of the place; and it is this harmony between nature and art, that is the greatest charm of these southern gardens; a sort of subtle, soothing influence that you feel, without being able to define; a vague impalpable unison from the drowsy perfume of the orange-blossom on the wall—a white spray of pearls set in the deep gloom of an ilex—to the moss-grown old faun, who, half drunken with its beauty, has fallen asleep in his niche below.

There is a fine statue, by Fedi, of the late Marchese, with a young figure by his side, the 'Genius of Charity,' in remembrance of his goodness to the people of Florence. In a season of great distress and poverty, he found work for many of the citizens in the construction of a high and curious tower, on part of his grounds, from the top of which one looks down into the heart of the city. We visited the hothouses, gay with flowers, and enjoyed the cool, shady walks.

The houses on the estate are let at high rents, two or three hundred a year, to English or Americans, and they are always eagerly secured by people who wish to enjoy a villa-like residence, and yet realise that they are within the walls of Florence.

We went together to Fedi's studio. It was too late to find him there, but we saw the marvellous group, his great triumph, 'Hecuba and Polyxena.' There is a talk of placing it in the Loggia de' Lanzi, where there is a vacant space; and all who have seen this most wonderful work must feel that it would hold no unworthy place by the great creations of past ages. It has been, or is to be, purchased by the municipality, on this condition, stipulated for by the sculptor, but there is still some prejudice to be overcome; some of the leading officials, who have a voice in the matter, being anxious that it should be placed in the centre of the Piazza della Independenza, the largest square in Florence. The group, though large, is far from being gigantic enough for such a site, where it would be dwarfed into insignificance. I hope it may find shelter under the stately Loggia, where the shadows of centuries, hidden in the dim corners, will sweep across it; and where just enough sunshine wanders in to light up the eager quivering faces. The city should be proud to feel it has at last a modern work worthy to be placed beside the statues of Cellini and John of Bologna. I wish I could make you understand its beauty, but no photographs have yet been taken of it, and my rough sketch is worse than nothing. Neoptolemus is carrying off the maiden to be sacrificed to the manes of Achilles; Hecuba in her imploring agony, a wild struggling woman, dragged along the ground, and her dead son, in his everlasting silence, lying stiff and still under their restless feet, while over him the three struggling forms in the passion and agony of life, hurl wild blows at each other, and cling and wrestle in the madness of their passion and pain. I never saw anything more wonderful than the woman who has fallen, yet holds on to the Greek with the grasp of despair, with wild, open eyes, and crushed and tumbled drapery dragging round her, showing the force with which she is being dashed along as she clings. In painting or sculpture, it is seldom that violent action can be well rendered, without its becoming an exaggeration, and at the best the eye wearies of the prolonged movement, and you long for the upraised hand to fall, or the strong muscular tension to cease, if but for a moment. I do not know whether you feel this as I do; that the works of inferior artists satisfy us by their harmony or repose; that it is only a great master who can dare to turn a moment of violent life into stone, and keep it there without startling us, by the mere power of its intense truth. Fedi, I think, does this beyond many; you feel that that mass of wild grief and fury might be whirled along, like Dante's passionate spirits, through ages to come, still battling and wrestling in unwearied wrath.

We walked back to the hotel, and then had an hour's drive outside the walls—the poor old walls, whose death-warrant has been made out, they say! Great alterations are planned, but the Florentines are slow to execute, at least there is not much private enterprise, but it is to be hoped the government will act promptly; so much is needed before the city will meet the requirements of a capital. The ministry are wisely adapting old buildings, suppressed religious houses, empty palazzi, to their different needs, without waiting to build, though new erections are planned, and wonderful changes and expansions, to meet present and future emergencies. Florence has no room to spare; its streets and houses are densely populated, and the walls necessarily hinder a gradual increase of the city proper; and so they are doomed, and one ought not to sigh over the loss of these old relics of a barbarous age. As we drove under them, I thought of the beautiful old city, rich and great and powerful as she had once been, but narrowed, self-contained, often eating out her own heart in silence, or destroying herself in useless struggles, within the shelter of that stone ring; while far away over the wide Campagna were many a fortress and little town, nests of sedition, where birds of prey fattened on each other and the poor helpless flocks in the valleys; and still there was the same blue heaven over all, but now a spirit had breathed over the people, making them one in a union earthly and yet divine. A ripple of sunshine laughed over the cornfields and the vines, and a long golden smile crept up to the rough walls, as if begging for an entrance, a murmurous whisper was passing among the reeds, and dying out in soft echoes against the old stones; and so the day is coming when Italy, free and strong, bids her cities open their arms, and let all happy outer influences pour into their hearts.

Florence, as everyone knows, extends along both banks of the Arno; what little commerce or activity there may be is concentrated at one end, where are the public buildings, markets, shops, and squares. It is here, on the north side of the river, that the greatest changes are contemplated. One or more new markets are to be built; the streets, where that is possible, are to be widened; there is a talk of a little public garden in the centre of the Piazza della Independenza; and, what is very much needed, fresh homes for the poor are to be erected, beginning with a number of small iron houses in an open space near the square of San Marco.

When the old walls are levelled, broad boulevards are to be made, running round the city, which will afford good frontage for the new houses which, no doubt, in time will form fresh rings round Florence. Towards the Cascine, the city has been extending itself during the last few years, large blocks of handsome buildings, stately modern palazzi, new hotels, broad streets and squares, have made a sort of 'West-end' to this side of Florence. Here is Madame F.'s beautiful villa, built by the Emperor, and the new houses of many of the Florentine nobility extending to the gates of the Cascine, near which is the old Pisan railway station, now converted into barracks. There is still available ground in this direction, and the city will no doubt spread rapidly, so as almost to surround, on its western side, the fields of the Cascine.

At present, there are few or no suburbs beyond the walls. The streets and buildings end abruptly, and, passing through the Porta Prato, you may plunge into lanes and drive between cornfields, and by old farmhouses, or the gardens of the villas that so thickly stud the valley and hills. On the opposite side of the Arno, where the Porta San Frediano opens upon the road to Pisa, there will probably be less of change; but as you pass back again, without the walls to the Porta Romana, and so again into the city, you drive under the rows of trees planted on the outside of the Torrigiani garden; and here again improvements are planned. A number of good houses or villas will be built by the Marchese, which, when the old city wall is thrown down, will enjoy light and air, and turn what is now a dismal thoroughfare of poor dwellings into a good suburb of Florence. The government have purchased the Palazzo Riccardi; and this and the Palazzo Vecchio will meet their most pressing need. There is some talk of the parliament assembling in the latter, and it is at present closed to the public, while the necessary alterations and improvements are being carried out. Murray describes the Riccardi palace as a species of Somerset House; it was bought by the Grand Duke from the Marquis. It has been long celebrated for its frescoes, its magnificent library, and as the meeting-place of the great Accademia della Crusca, and also for its architecture—the lower windows (by Michael Angelo) being the first example of a sill supported by consoles. In the walls are, as usual, the great iron rings, in which the retainers set their flaming torches.

The great marvel to me is, how the middle-class Florentine exists. There is no visible trade or manufacture beyond the copying of pictures, and the great mosaic and china works, which may be classed together under the latter head. The women and children in the villages plait straw; and, in driving through the narrow back streets, you see artisans busy over their various occupations, labouring to supply the necessity of the hour; but apparently there are no merchants, large or small, no capitalists engaged in business, no 'city men' who might have worked their way up to independence, and carried with them the vigour of the people, to be absorbed with time into the classes above them. There is no healthy *growth* amongst these Florentines; the people all seem to live on a fixed income, and never to strive to increase it. And yet the success of the lottery proves that there is a sufficient amount of lazy love of speculation in the people, if they could be only sufficiently stirred up to turn the wheel for themselves. You hear sometimes of a Conte or a Marchese who has established a manufactory of pottery, or of a successful speculator in copper-mines or marble quarries; but these latter, for the most part, seem to be strangers. In everything, Florence is far behind the more northern cities of Italy. The Turinese have inherited more of the active sturdy spirit of their ancestors, and they are flocking south to put their shoulders to the wheel. The Florentines repulse them jealously, and the workmen sent by the government to assist in the new buildings were not made over-welcome, though the stories of their ill-treatment seem to have been somewhat exaggerated. The few really good shops that one sees have all been opened by strangers; a Milanese draper, an English chemist, and baker, a French milliner, and so on. It is to be hoped they will introduce a fresh system, a *prix fixe* being sorely needed in a place where great ladies will chaffer over the merest trifle; the shopkeeper, as at a Dutch auction, starting by demanding a sum six times its value.

New markets, good shops, broad streets, will make a wonderful change for the better; and the infusion of a little fresh blood and a new spirit of enterprise may strengthen and rouse this quiet, placid, lethargic people.

Good schools for the children of rich and poor—for both, I fear, are equally ignorant—and anything that will stir them up to self-improvement and a healthy emulation, will work wonders for the Tuscans, who have so long lived with very little need for thought or action, crushed by the priests, downtrodden by the government, that any sense of a desire for what is better beyond a vague yearning for liberty in the abstract has been ground out of them.

There is a vast amount of charity in Florence, public and private; the doctrine of its covering a multitude of sins being received by priest and people as applying to the *bestower* of the charity! Even the very poor have their pensioners, and a day in each week will be set aside when those more utterly destitute than themselves come for rags or bread, or the little drop of oil that is such a cherished ingredient in the soup. The religious houses have, as of old, their flocks of retainers, idle or poverty-stricken or lazy, who subsist on alms.

Fortunately, mendicity is forbidden within the walls; but there is a swarm of lame, halt, and blind, waiting the arrival of the unwary traveller on every hill near the gates. They crowd round the carriage, when the steep ascent places you fairly at their mercy, whining out their demands for aid, '*Per amor' di Dio! Una piccola moneta, per amor' della Santissima Vergine!*' and ready with muttered maledictions, when, having reached the crest of the hill, your driver whips up his horses, and they are forced to let go their hold, as you dash round the corner, glad to breathe a purer atmosphere, not infected by their blighting presence. Many of these people, the regular specimen beggar with a 'show' stump, or some fearful and much-prized disease disfiguring face or limbs, eagerly exhibited for compassion and *quattrini*, must have large and regular incomes, and can probably accurately calculate the amount which each of their patrons may be supposed to represent. Many of the Italian nobles regularly bestow a certain sum of money daily in this promiscuous charity; I have only heard of one who, while famed for his princely generosity to the poor, could say, when dying, that he had never given a centesimo to a beggar.

I was told a story the other day, which I believe to be only a fair illustration of the angry spirit with which these unfortunate miscreants demand alms at the roadside, though one would hope, for the credit of human nature, the *act* was an isolated one.

A woman, apparently miserably poor, with a sick child in her arms, had attacked some ladies who were driving in a carriage, and finding her petitions unheeded, she flung her infant into their arms, exclaiming, with a fearful imprecation, that her little one was ill with the small-pox, and she prayed Heaven to send the malady to 'these hard-hearted ones.' It was only too true; the fearful disease was raging at the time, and, whether from contact with the child, or terror at the woman's words, I know not, but one of the ladies speedily sickened and died.

May 24.

We have had lovely weather since the heavy rains two days ago, a cloudless sky and pleasant westerly wind to temper the heat, which has been very great in the middle of the day. I stayed in, as usual, all yesterday till the evening, when we had a charming drive in the country, passing out of the city by the Porta Prato, on by the magnificent Villa Demidoff, closed during this month, and so round by winding lanes to Fiesole and the mountains, the high hedges on either side hiding the corn, fruit trees peeping above them, and masses of wild roses and honeysuckle hanging down towards the road.

The scene was very English, if one could fancy the vine leaves and tendrils were part of a Kentish hop field. The heavy rains had laid the dust, and the roads were good. Large farmhouses rose at our side every now and then, and little clusters of houses, 'parishes,' as the villages are called, an old church with a quaint beautiful tower watching over them, and shrines of the Madonna, bright

At the gate of the Cemetery.

Hiram Powers.

Ascension Day.

with little fresh bouquets. There were groups of children and women at every doorstep, knitting or plaiting straw; cows and goats just milked and driven home for shelter; workmen returning from their day's task, which lasts till seven, with an hour or two's *siesta* in the great heat.

It was strange, at a sudden turn in the Devonshire-like lanes, to see the great tower of the Palazzo Vecchio at the end of its windings, or watch the domes of Florence in the distance against the soft purple hills. We skirted the city, entering again at the Porta San Gallo, after enjoying to the full the fresh beauty of everything around us. The perfect repose, so near the heart of a great city, had in itself a rare charm. In the fields were men busy stripping the leaves from the mulberry trees, seated on the branches, or climbing ladders; and the newly-cut hay scented the air.

The farmhouses generally bore the arms and crest of the proprietors; they are always inhabited by a *tenant*, who works the land, paying to the owner three-fourths of the produce of the farm; oil, wine, wheat, maize, all are equally divided, the landlord on his side assisting in the purchase of stock and seed. This system seems to act badly for the holders of small tracts of land, though of course they are infinitely better off than our agricultural labourers; it keeps them from the actual failure that might lead to their ejection and the absorption of their land by larger tenants; but they may be ground down by an avaricious landlord, who can claim his share even of the daily return of eggs! Everything is managed by the *fattore*, or steward, who renders what account he pleases to his master; and as the owner can have little or no power of testing his accuracy, the servant has a hundred ways of enriching himself at his expense. '*Fatemi fattore per un anno; se son' povero è il mio danno,*'—'Make me a steward for a year, and it is my fault if I am poor,' is a common Italian proverb. These men often acquire large fortunes, and are able gradually to buy up the lands of the needy noble whose property they have consumed for so many years, and thus there rises up a sort of third estate, middle men, who slowly but surely seize upon the land, and it is to be hoped, profiting by their own experience, introduce a better system on the farms they have acquired.

So far, all attempts to change the old farming institutions have failed. An Englishman, who had purchased a large estate near the city, determined to keep it in his own hands, but the difficulties to be encountered were so many and so insurmountable, that he had to relinquish his experiment, and yield to the received habits of the country, and be honestly served or pilfered by a steward and tenants, as the case might be.

The amount of wine made in Tuscany during the last few years has been much less than in the times of the famous vintages, when the choicest juices seem to have flowed like water, and well-filled flasks were to be had for the asking. Since the grape disease came as a terrible foe to the poor peasants, the quality, as well as the quantity, of the wine has been affected, though the country immediately around Florence, and the more distant hills and valleys, still produce many famous wines, red and white, of wonderful purity and richness; the *Montepulciano*, the *Aleatico*, *Chianti*, and many others. Notwithstanding the failure of many years' crops, the wine is still sold for what to us would seem a wonderfully small sum; but the landlords of hotels take good care to increase its cost by the addition of a considerable percentage, employing that delicately graduated scale of prices, common even to those innkeepers of old times, who drew any wines that might be called for indiscriminately from the same cask.

Not that the wine in the hotels at Florence is adulterated, except in as far as it is toned down with copious infusions of a purer element. It is amusing to watch the gradual paling of the red wine served at our table d'hôte, and on the arrival of Mr. Cook and his six-and-twenty companions, the carafes looked as though they had been filled with water, with a slight infusion of pickled red cabbage. On my father remonstrating, the secretary assured him he should be supplied with some wine recognizable as such. As I hardly ever drink any, it seemed a pity to have my allotted portion of water added daily to the modicum of grape juice served out to us.

The people eat many vegetables as *salad* which we only consider wholesome when properly boiled; peas and beans are constantly served up by them, half raw and quite hard, with the addition of a little oil and vinegar. The dish of asparagus which was sent up to C.'s room yesterday was utterly uneatable, much to my annoyance, as it was too late for the cook to procure a fresh supply from the market, and a second boiling seems to have no effect, just as people declare it is impossible to alter the state of an egg that has been once removed from the hot water.

We handed over our asparagus to Francesco, who, with true philosophy, at once appropriated the dish to his own use, turning it into a salad and eating it cold 'to the health of the Signorina,' as he informed nurse. She gives him lessons in English, which, like all waiters, he is very anxious to acquire, and he is supposed to keep a notebook, in which he carefully enters any new word he has learnt; and it was with great secret amusement we listened to Carolina, who gravely informed us that poor Francesco had just written down '*sparrow-grass*' with great glee as the name of his last salad!

How oddly inaccuracies are perpetuated, and with what perfect good faith people talk the patois of other countries! It reminds me of the Polish lady, who said to an Englishwoman—

'Madam, after the greatest exertion, I am at length repaid; I flatter myself, I have secured a governess for my children who speaks the *purest cockney*!'

XVII.

May 24: evening.

We saw, one morning lately, the 'Spedale di Santa Maria degli Innocenti, the Hospital of the Innocents, a handsome old building in the Piazza della Santa Annunziata. It is long and low, as are most of these institutions in North Italy, with an overhanging upper story, supported on pillars, the outer wall adorned with Robbia ware, a little white papoose on a blue ground in a medallion between the arches. The inmates are very young innocents, babies of a few hours old, who are deposited overnight in the open drawer or receptacle provided for them, and drafted off the next morning to the country, where they are reared by hired nurses in the many little villages hidden away amongst the vines and olives. The institution is highly valued by the poor of Florence, who make large demands on its aid. Carolina told us that last year she was hastily summoned to one of the houses in the street in which she lives, Via Benedetta, and willingly went with the messenger to render any help in her power to her poor neighbour, whom she found very ill and destitute, with her new-born infant lying beside her. Nurse did all she could to help her, and then inquired, 'What are you going to do with the child?'

'Oh, he is to go to the 'Spedale; my sister will carry him there presently; but I have not even a rag to wrap him in; if you will have the charity to give me an old piece of flannel, all the rest will be easily arranged.'

Carolina said the woman was the wife of an artisan earning regular wages, and though poor, they were by no means really destitute; but the habit of self-dependence had been so utterly lost, that year after year the mother was accustomed to dispose of her young babies by throwing them on the care of strangers, not even troubling herself to prepare any garments for them. She said she had had three or four children, and always sent them to the hospital, where they were well nursed, and when they were five or six years of age, she reclaimed them. 'She could not manage with little babies; children wanted so much looking after.'

But, we asked: 'How do they know their own children?'

'Oh, they tie some ribbon round their necks, or cut a centesimo in half keeping one of the pieces and fastening the other on the child, or stitching something to the clothes, when they have any; and so they fancy they can find them again. They get *some* child given back to them, but whether it is theirs'—with a shrug—'bah! who knows?'

When the infant belongs to people in a higher rank of life, who have placed it in the hospital probably to conceal its birth, more care is taken; some one is sent to the building early the following morning, as the country nurses are in waiting each day before ten to receive their fresh charges; the nurse to whom the baby has been given is marked and followed, and often the real address of the mother entrusted to her; and they are, of course, always glad to know the parents of the child, as any extra care or tenderness bestowed on the poor little mortal is sure to be rewarded. I believe these children are really well cared for, and probably grow up a finer race physically than if they had struggled through their first years of existence in the dark, narrow streets of the city; but the effect of such a system on the family life of the poor must be very bad. The terrible poverty and distress of many English cottage homes makes one hesitate to condemn a custom, which, as Carolina's poor woman explained, left them free to labour for them in the future; but surely, nothing which severs the tie of dependence which God has implanted, if only as an *instinct* in man and beast alike,— the hourly need of loving ministrations given and received, the devotion of a mother to her little children, which is often so heroic in the patient tender toiling of the poor,— can be justified by any state law, however humanely conceived. You must remember that there is no disgrace attached to these foundlings; it is the custom. The Government is benevolent, and its infant population is put out to nurse at its expense. The hospital was established as early as 1421, and receives annually above 3,000 children from all parts of Tuscany; the Grand Duke was considered, in the eye of the law, as the father of these very young subjects, and, until they attained their majority, if any of the children got into trouble or were charged with theft or any other offence before the tribunals, he was expected to send a representative as a sort of counsel for the prisoner. The Grand Dukes are praised by the people for their benevolence, and under their rule many of the laws seem to have been well executed. Their internal policy has always been to satisfy the people with shows and charities, to amuse them with numerous holidays and fêtes; the Church going hand in hand with the Government in striving to keep them as children, ignorant and content, less able to think or act for themselves the more this paternal government thought and acted for them. They are a handsome people, these Florentines, and look handsomer than they are; the women are well grown, tall and slight; they all walk well and carry their heads gracefully; in passing through the streets nothing struck us more than this contrast in their appearance to Englishwomen of the same class. Their graceful step and very quiet manner give the plainest woman an air of good breeding; they are thoroughly ladylike, and their simple, tasteful dress sets them off to the utmost advantage.

A large party of travelling English were expected here last night, one of Mr. Cook's great excursions. He has been three times before to Florence, and brought eighty people to the city last year! There are only thirty in the present party; they will only stay three days, and we are looking forward, with much interest and some amusement, to the prospect of encountering them at the table d'hôte.

May 25.

I shall have more time than I thought to find, to finish my letter in to-day. Carolina called me at the wrong hour, so I am dressed by seven instead of eight, and she, poor thing, roused herself before it was necessary. Ascension-day is one of the great fête-days of the Florentines, when all the city hastens to enjoy out-of-door life in the Cascine. Rich and poor alike make a joyous pic-nic. By four o'clock crowds of people are in the fields breakfasting under the shade of the trees, or on the open grass, carrying their bread or cakes with them, and buying milk from the great dairy. As I write, these early excursionists are returning along the Lung' Arno; neatly dressed groups of the very poor, with napkins folded in their hands, or empty baskets, all that remains of their Cascine *colazione*; carriages are flying backwards and forwards; little family groups pass the window, preparing for a day's holiday, the provisions tied up in a handkerchief carried by the father or the biggest boy; some there are so exceedingly grand and superior that their commissariat has been sent on ahead, and the young girls have nothing to do but to guard their dainty muslins from the dust, or hold up the large paper fans when the morning sun strikes too powerfully on their uncovered heads. The very poorest wear fresh, delicately-coloured dresses, with jaunty little aprons, their black hair put back into the long Spanish-like net and dressed high over the forehead. The next step in the social scale is marked by the muslin being longer and clearer, and by a bonnet.

French fashions everywhere are driving out the quaint old costumes which were once so picturesque; the people themselves are growing ashamed of them. Carolina spoke to us the other day, with great contempt, of a lady who allowed her nursemaid to go about with her wearing a handkerchief tied over her hair—'it is only the *poveraglia* who do that.' Here and there you see a contadina, with a string of rare pearls, or the necklace that her great-grandmother brought as her dowry—its great pink coral beads nestling against the soft brown neck. Still in the markets you may watch the countrywomen unloading their patient mules, sheltered beneath their broad flapping straw hats, that have seen many a summer's sun, as the hard-working wives and mothers, with a child in their arms or the busy distaff in their hand, tramped along the hot white roads from Careggi or Fiesole, through the Porta San Gallo; but these hats, once almost universal, are rapidly diminishing in numbers, if not in size.

The Tuscans are great straw workers. In driving round Florence, you see women and children sitting at their doors, all busily plaiting the straw, and in many shops there is a large display of very beautiful work—birds and flowers, and fanciful braids and tassels.

We see very much that is amusing from these windows, sadly too sunny ones during the heat of the day; then all the jalousies have to be carefully closed, and as there are plenty of doors to make a thorough draught, the rooms are pleasant enough, with their shady green light, and the dreamily scented air from the great bouquets of magnolias and orange blossom and Cape jessamine, with which I delight to fill them. Early in the morning we are roused by bursts of music, as the soldiers pass to their parade, and the trumpets of the Bersaglieri always call me to the window. I like to watch them rushing by; the men, small, active, wiry, well-developed fellows, in their light summer undress, and dark belts and jaunty hats, the long feathers streaming in the air. It almost takes away one's breath to watch them coming down the street, and the sort of swing trot with which the band shake the music out of their instruments. I am afraid the service is a hard one; they say the men do not remain many years in the regiments, and that they are permanently injured by the pace at which they are made to walk. It does seem cruel to keep it up, when one sees them coming back from parade, hot and tired, and the sun has risen higher, and the pavement beneath is beginning to glow again. But the sound of their trumpet always stirs one's heart, like the old Garibaldi hymn, and the men have such bright, glad faces, and look as though they could sweep all before them, so that I hope the over-walking is a myth, and that the pace is not so killing, after all.

When the soldiers are gone, then come the morning loungers, and a little crowd of citizens, not over many, for this is not a commercial part of the city, and there is another street behind the hotel, by which the supplies for the houses can be brought, and in which we have discovered an English baker, from whom we buy good biscuits, which no foreigner, strangely enough, ever seems able to manufacture. Men are busy in great wooden barges, digging up the sand from the river, which after any sudden rain is brought down in great quantities from the hills, dyeing the water a deep yellow brown. It is valuable to the builders, who use it for making mortar, and the men look picturesque enough standing cooling their bare legs in the stream, and chattering and singing in chorus. As the sun rises higher in the heavens, and the edge of shadow on the pavement and beside the wall of the river shrinks into nothingness, the street grows more empty, and a great silence falls upon the place. The only living thing to be seen from the windows is a man lying against his barrow and fast asleep, browning in the sunshine; or a dog who wakes to snap at the flies, and lazily shake himself, before stretching out for another doze on the hot pavement.

About four o'clock the world begins to move again, carriages hurry past, and by five or six there is a gay crowd on the Lung' Arno, and a steady roar of wheels, as all the rank and fashion of Florence whirl by to the Cascine. As the evening goes on, the people crowd towards the river for air; hundreds of men, women, and children stroll along the pavement, or sit chatting in the small piazza by the Ponte alla Carraia, men smoke, leaning against the parapet, and there are always rows of boys perched on the wall the whole length of the Arno, their legs idly swinging in the air. A good-tempered, inoffensive crowd, finding enjoyment in mere existence, talking in soft, low voices in the liquid syllables of the *lingua Toscana*, regardless of any pleasure or excitement beyond the amusement of watching the gay occupants of the grander carriages, or perhaps venturing on the extravagance of a halfhour in a *vettura* themselves, but this generally on a Sunday, when the small savings of the week are devoted to a drive in the Cascine, the whole family being stowed away inside or on the box, while the little carriage overflows with laughter and happiness.

The people are mild and gentle, with few apparent vices, and seem born without the eager longing for excitement which stirs so many city crowds in England. The lottery works off a little surplus restlessness sometimes, but I fancy the picture of its horrors has been overdrawn; it may have made some victims, but it is supported in the main by small ventures, people putting in their money occasionally more as a *passe-temps*, just to try their luck, than as the habitual gamblers we had been taught to consider them. Of course it is an unmitigated evil; but even regarding it as poisonous dram-drinking, I do not think there are many who are constantly taking a dose. The people are essentially moderate in their habits and pleasures, very frugal, and easily satisfied. Their food is of the simplest; a cabbage boiled in water, and a little sweet oil, makes a soup that would satisfy others than the very poor. They seldom drink anything stronger than coarse coffee, or the weakest wine of the country, and there are literally no places of entertainment, like the tea-gardens and beer-houses of England, round Florence. When work is over, the people sit out of doors, or walk to the Cascine, or to a shady plot of ground near the Porta San Gallo. As some one once said to us: 'The air is so exhilarating, they do not require stimulants.' Throughout all the excitement of the three days of the fête, we did not see a single person in the slightest degree intoxicated, and the streets were always quiet and silent by eleven or twelve at night.

The Florentines are very fond of a siesta, even quite poor people going to bed regularly of an afternoon. The ladies, being untroubled with morning callers, often live in a demi-toilette till they dress for their dinner and drive. The nobility are in general too poor to give dinners; an occasional musical party, or a ball during the Carnival, is the extent of their hospitalities. There are, of course, a large number of English residents who visit among themselves, and it is wonderful what odd glimpses we get into the different 'sets' of society here from our acquaintances.

Some of the English are utterly unprincipled; thoroughly bad English people abroad being generally, I fear, rather more wicked than the worst foreigners they settle amongst. The increased price of everything will drive away the tribe of miserable adventurers who, for so many years, have hung about the city and its neighbourhood, leaving a number of pleasant, highly-educated families who have made Florence their home, and have helped to make England loved and respected by the Tuscans.

There is the 'fast' set, the literary, the fashionable, the high and low church, the sociable, the exclusive, the evangelising, and numbers, too, of those useful members of society who fit into odd corners, and are round or angular according to circumstances. Each of these divisions is subdivided *ad infinitum*, and if the bad people quarrel, the good ones do not, perhaps, love each other as much as might be wished. There are a great many Americans settled here, and naturalised by residence or marriage; they form a small coterie amongst themselves, and through last winter have met once a week in an old palazzo across the river, where a room has been fitted up as a tiny theatre for private theatricals.

An American girl, who sat next C. at dinner one day, was telling her of these performances. She went on to describe her recent visit to England, saying she had been in London for some weeks, and, much to her disappointment, had never seen a fog. All her friends who had returned from *Eu*-rope after visiting London had seen a fog! She had evidently regarded it as a regular sight like the Thames Tunnel, and felt cheated. C. asked whether she did not admire the beautiful English hedges.

'Why, yes, they're handsome enough, but I ca'c'late we ran fast enough to see a stone wall at the end of a fort-night!'

It is not very difficult to obtain an *entrée* into very good and really pleasant society here, but we originally intended our stay to be so short that we declined all invitations or introductions, and when it was prolonged by C.'s illness, we had, of course, neither time nor inclination for anything of the sort, though we value the few real friends we have found here. I have run, as usual, into parentheses, and wandered away from my window, but there are always so many little things I want to tell you that will

Grillo.

Ascension Day.

In the lanes.

come into my head at the wrong time. You must make the best of such an *olla podrida*; the ingredients are very genuine, though I cannot make them piquant. We meant to have bestowed upon you such interesting and erudite epistles, which would have assisted you in forming your daughter's mind, and now Totty, who has already arrived at the mysteries of *patchwork*, will consider herself as accomplished as her aunts! We never aspired to 'fine writing,' but we had hoped to carry you in some degree along with us, and to make you realise in what a dream of beauty we have been living, or more truly, that we have awaked to find all we ever dreamt of a wonderful reality, that ought to fill one's heart with good and great thoughts, to be stored up for after years as an ever present joy; and week after week you are spelling instead through pages of gossip, the small facts and stories and sights of our daily life and the life around us.

Small boys all through the day have been shouting, 'Grillo! grillo! grillino!' in the streets, while they carried, hung on a stick across their shoulder, tiny cages of straw or basket work, about two inches square. This is as much a part of Ascension-day in Florence as the mass in the churches, though what it means, or what is the origin of the custom, who can tell? My father has just come in, with one of these little cages in his hand. The insect is a kind of cricket, and the people keep them imprisoned all day, that they may hear them chirp at night, and find perhaps a promise of good fortune in the sound.

The carriages, as they drive by, are full of little children, to whom their pretty dark-eyed mothers seem tender enough. While they wait in the Piazzone, the little ones and their nurses are playing in the soft grass of the Cascine. These *balie*, or nurses, still wear a costume like the Albano women, who, in their bright-tinted dresses, watch over their little Roman nurslings. They look handsome and picturesque, with lace lappets in their dark hair, long white aprons, and broad gay-coloured ribbons, blue, or yellow, or scarlet, on either shoulder.

XVIII.

May 25: evening.

I HAD my painting lesson again this morning, but I can do little after breakfast, as, even if C. were well enough to spare me, the sun shines too strongly on our rooms; the jalousies must be kept shut, and I am not sufficiently experienced to venture on delicate gradations of colour under a strong *green* light; and as our rooms are all front ones, there is no variety of aspect. This afternoon I did attempt a little, as I was anxious to finish the second painting before Signor Pompignoli's arrival tomorrow; but I was exhausted with my efforts; the heat was so intense that the weight of a maul-stick seemed too much for one's powers, and everything went wrong, and the paint dried too rapidly and was sticky in places, and my canvas and mind both seemed rubbed the wrong way, and there was a plague of flies who buzzed round the ends of the brushes and settled on one's face, till at last, patience being at an end, I made a sudden *razzia* on my tormentors with palette, brushes, maul-stick, and all, upset the oil, painted myself all over with burnt sienna, and determined never again to make such a hopeless attempt, but confine my work to the two hours before breakfast. That daily bit of good steady work helps me *morally* through the day, and makes C. feel it is not quite lost when we spend the rest of it quietly in her room; and now that she is a little better, I can often have an hour in the galleries or churches with my father. . . .

There is so much of constant interest and amusement going on around us that we are never dull; with a good supply of English books (the last packet of magazines from home was most welcome), work and talk, pictures and flowers, it would be strange, indeed, if time hung heavily on our hands; and now C. has ceased to suffer so much, we are very thankful and happy together. * * *

The table d'hôte was crowded with Mr. Cook's party; a very curious one, young and old, parents with their sons or daughters, single men, young clerks and governesses, and elderly females. They arrived at two in the night, or rather morning, and there was a considerable scrimmage, one of the waiters tells Carolina, before they could all be stowed away in their respective rooms, two fat old ladies appropriating the one they considered the best, and, to the servants' great amusement, at once removing their dresses, considering possession to be nine points of the law. I regret to add they were turned out ignominiously in the end. Our pleasant German neighbours sat near us again, not the *madonna*, but a little brunette and her husband.

The evening before, Signor A. called at half-past five, when we secured a carriage with good horses, and drove up to the hills. The expedition was a great success, and the fresh pure air delicious, after a busy day in the sick room. Passing through the Porta Romana, we ascended by the same road we had followed in visiting San Miniato, only going much farther, getting lovely peeps at every turn of valley and city, passing the house of Galileo, with his bust in a niche in the wall, festooned with laurels, and the tower, misnamed Galileo's, the *torre del gallo*, so called from the vane on its summit. From Galileo's villa one commands a glorious view of city and valley, and the grand sweep of the amphitheatre-like hills; and there how often must he have stood, surrounded by the few chosen friends who were faithful to him through all, working out noble truths which were to live when he had crumbled into dust; here, close to his beloved city, yet an exile from its walls, he could feast his eyes with its loveliness and its memories, and rejoice in the still Italian nights, when the heavens in their solemn repose smiled down at him through their thousand starry eyes that he had learnt to know and love; and when a deeper darkness closed upon him, the great thinker, blind and feeble, must have sat there calmed by the soft evening breeze, his soul wandering away into the immensity of space, into the eternal hereafter, when all truths should be made known.

We stopped on some rising ground crowned by the little church of Santa Margarita, with its *curato's* house and garden, such a pretty garden! with a little terrace and shady arcades, and a table and chairs for the after-dinner coffee, a rockery, and flowers and ferns, and a boarded end, painted like a scene at a theatre, to give unlimited idea of space, combined with snugness and seclusion. There was an honest old gardener, who invited us to ascend the belvedere, and busied himself in preparing a bouquet for me, looking very much distressed at being requested to accept a small *dono* in return.

We sat on the little terrace and thought the curé was a happy man among his brethren; mere life in such a place must be a delight; no words can describe the beauty of the scene, and I have already expended all I possess in writing to you of similar ones, only that each has its own particular charm. I have great sympathy for that much-suffering school girl, who, when desired to write a piece of composition, a description of an English landscape, threw down her pen, declaring with dismay that the governess had forbidden alliteration, and she had already used every adjective in the English language in the first page of her essay!

The evening was perfect; we looked down on the city, a little hidden by a great shadowy mass of hill, which we had passed beyond. The more distant ranges were still bright in the sunshine, and the plain a golden green; corn and fruit trees and creeping vines in rich luxuriance, sloped away from our feet. As we watched and rested, the vesper bell rang out from the tower, and another, and another, softly tinkling over the hills, took up the echo, and the tender little call to prayer and thanksgiving died away in a musical murmur, the chirrup of the cicale below us, and a sweet thrilling nightingale in a tree above making instant answer.

But the cypress shadows were lengthening warningly, and we rose to go, exchanging a cordial *buona sera* with the pleasant-faced gardener, who was bringing water to the thirsty flowers, with may be also a *benedicite* in his heart.

On leaving the garden we looked down, on the opposite side of the church which crowns the hill, over another wide-spreading valley, watered by a bright little stream, the Ema, winding amongst the trees, and dotted with villas, and far away, the road to Arezzo, disappearing in the distance, leading to Rome!

We drove round to the Villa Fenzi, entering the city by the Porta San Nicolo, and facing a flood of orange and crimson light as we passed the end of the Uffizi, and drove down the narrow street leading to the Lung' Arno, the high buildings on either side flushed with the sunlight or deep in shade, the sky before us one golden glow, a great vermilion sun sending down splashes of red into the tawny waters of the Arno, and the delicate outline of the Carrara hills, drawn with a serrated line of white fire against a bank of violet clouds. Over our heads the blue sky was paling with a green glimmer where it looked towards the sun, lamps were twinkling in a brightening row along the street beside the river, and the bats were fluttering above them, or hurrying in their whirling flight from cornice and window of grim old Palazzi, emerging suddenly where the shadows were deepest, like unquiet spirits, black, and bad, and restless.

Slowly a golden moon rose into the heavens, the long lines of colour died away in the west, the stars came out and the blue vault veiled itself in a soft purple mist, a gentle dew fell upon the weary earth, and, slowly turning away my eyes from the beauty of the scene, I joined C. in her room, and tried to picture it all over again to her, while we shared our warm milk and strawberries and little sponge-cakes together.

May 26.

* * * To-day, after a busy morning, callers, letters, the doctor's visit, &c., my father and I drove to Santa Croce to see the grand Dante statue, a wonderfully fine work; it is dignified and full of expression, a noble disdain on the face and figure, rather than the sour disappointment too often written on his face. The pedestal is not yet finished, but the bas reliefs were cleverly sketched out, some fine compositions from his great poem: four lions are at the base.

From thence we went to the Casa Buonarotti, and spent some time examining the Michael Angelo memorials. I think I told you that the last of the family married an Englishwoman, a Mrs. Grant, and she it was who arranged and prepared all these rooms with appropriate frescoes and paintings. We saw the sword, and walking-stick, and Eastern slippers, and some of the great painter's writing, a sketch in oils of a Holy Family (larger than life), studies for statues and bas reliefs, and the original sketch of part of the Sistine Chapel paintings. Signor Buonarotti died a few years ago, and his wife did not long survive him, but she was anxious to the last to preserve for ever the memory of her husband's illustrious ancestor; the house and collection were left to the city, with an endowment for their preservation.

We drove next to the Bargello, where we only stayed a very short time, visiting Giotto's fresco, which disappointed me. Dante's face is much less beautiful than in the Arundel reproduction, and the mass of indistinct heads surrounding it lessens the purity and clearness of the outline. The painting must have been re-touched to a very great extent, or the contrast it presents to other figures in the procession would be most unaccountable. We met the C.'s again at the Bargello, and visited the Mediæval Museum together, returning by the Mercato Vecchio, to lay in a good store of strawberries. Now that we can be depended on as customers, the people keep a daily supply of English strawberries for us.

At dinner to-day we found none but Mr. Cook's party; a gentleman, his wife and two daughters, worthy Londoners, who sat next us, told us a great deal about their proceedings. They had intended spending a little time in Switzerland, but not having much experience in the mysteries of 'foreign travel,' some one had recommended them to apply to Cook. They had first-class tickets, an arrangement which is optional, and had come with the rest of the party, viâ Neufchâtel, the St. Gotthard, and the Italian lakes, to Florence for *two days*, and are to start at five to-morrow morning for South Italy, by rail and diligence, stopping one night in Rome, and going on to Naples, all by land; returning again to Rome for a few days' visit, and so to Leghorn by water, and by Genoa and Mont Cenis to Paris, the whole journey to be accomplished in one month, at an expense of about twenty-five pounds a head—*first class!*

It is a great pity Mr. Cook should attempt to take a crowd of people scampering through Europe at such a pace; they can enjoy nothing properly, and will probably return with an idea that they have seen everything, whilst their real knowledge and appreciation of art and nature must necessarily be very limited. One is almost tempted to think Cornelius O'Dowd's criticisms not altogether undeserved. The Scotch, and Irish, and Swiss journeys used to be well managed; but this is, I think, a great mistake.

They all seemed on very sociable terms. We asked how they managed to avoid squabbles about rooms, and places in carriages, &c. There had evidently been some disputes about the *coupés* of the diligences over the St. Gotthard, for in future Mr. Cook decides to charge five francs extra for such seats, to avoid the rush of claimants. In writing to the hotels, he orders 'so many rooms for a family of four, so many for three, or one, or five,' and then each family is shown to their rooms on arriving. They have one large sitting-room in common, all their meals being served, I fancy, in the *salle à manger*; and last night, on their return from the Cascine, where they had been sent in numerous little carriages, they had an amusing evening together, and were, they assured us, 'quite lively' over music and dancing. The oddest thing is that Mr. Cook himself cannot speak a word of any language but his own.

The poor creatures have requested to be called to-morrow at three. Fancy such a journey, in such heat, for delicate women and girls! and they will reach Rome and Naples just as everyone is leaving, or rather has left, for the continuous stream of travellers has ceased to flow north. My father warned Mr. Cook of the Roman fever, of which he knew nothing; and yet he professed to be able to care for all who were travelling under his directions. They run a frightful risk, encountering the air of cities that have been far from healthy all the winter, weakened by a hasty journey, and venturing south when the great heat has driven away all their countrymen; the journey, of course, being planned out of the season to insure cheap and plentiful accommodation for the party at hotels, where they are made very welcome.

But to return to my journalising. Soon after five o'clock we drove towards Fiesole, Signor A. accompanying us. Just three weeks before, we were there with C., and we hardly dared to realise all the deep anxiety we had passed through since our happy evening on the hill.

Instead of ascending to Mr. Spence's villa, or above it, we drove on along the high road to Arezzo; the more distant views lengthening out before us, and Florence nestling in the valley on our right. We very much enjoy exploring the environs of the city. It is curious how many pass through Florence without visiting the hills at all, neglecting even the favourite drives to Bellosguardo and San Miniato to which Murray directs their attention. This is partly owing to the coachmen's extreme unwillingness to go anywhere but to the Cascine. Carriage hire is still moderate, though, like everything else, it has nearly doubled in the last few years. You always pay less for the second hour, and in the Cascine the carriages are generally drawn up in the Piazzone for a third of the time, an arrangement naturally agreeable to the horses. The hill roads, though very steep, are generally good, and the horses are strong and able; but as the men get little more for a distant excursion than for the quiet *giro* in the Cascine, they are always eager to go there, as during the long summer evenings they can make two or three engagements between the dinner hour and darkness. Often we have waited till our patience was worn out, endeavouring to find a man willing to take us an ordinary drive, those belonging to the unnumbered carriages asking extravagant prices, while most flatly declined to go with us. Signor A., who has lived for so many years in Florence, and has explored every nook among the hills, is a capital cicerone; when he is with us we take long drives into the country; or, if my father and I are alone, we combine a church or museum with an hour by the Arno.

It was amusing enough in the Piazzone the other night, when we happened to have many acquaintances there, and drew up our carriages together, with much pleasant talk, to break the monotony of the drive under the trees. The light wicker or iron-work carriages are very much used: a sort of calèche, very light and elegant; there are a few pony-carriages driven by ladies; two or three women on horseback, principally English; an English-looking four-horse drag, belonging to a Florentine noble, who, however, seems generally to leave the reins to his servants; a throng of well-appointed equipages—some with a *chasseur*, gorgeous in green and silver and waving plume—side-by-side with a *rettura* from the stand in the Piazza S. Trinità, with six lively citizens, parents and children, talking, laughing, gesticulating, while the small patient little horse stands on three legs, wisely resting a lame foot and dozing quietly, his master fast asleep lying half across the seat above him. Near by, a military band is uttering sounds anything but musical, to the great delight of an eager crowd around it. Sometimes the soldiers burst into song; their rich resonant voices drowning the more discordant instruments. Little vehicles, like a sort of gig hung low between the large wheels, dash in and out, the small pony always at a canter, except when whip and shouts have urged it into a gallop, long tassels of wool or hair streaming from its head and harness. These are owned by the artizans or shopkeepers of the town who delight in their evening drive, the little carriages being useful in many ways during the earlier part of the day in their various avocations. The unfortunate ponies must have rather a bad time of it!

May 28.

* * * * The weather has become, if possible, more 'summer-like' than ever; but we arrange our days

In the Piazzone.

Ponte alla Carraia.

accordingly, and so do not suffer from the heat. I am wonderfully well; but then, though every hour is so busy, I use very little exertion. This morning, for instance, it would have given me a *coup de soleil* to walk to church, but being driven quietly through the air, it is not very oppressive; and the coachmen always take you through shady back streets. * * * *

The doctor, too, has proposed change of rooms, but C. begs to remain where she is; she is fond of our pleasant quarters and her view over the Arno, and likes the amusement of being so near the street and river, and hearing what is going on, and the noise does not trouble her now she is a little stronger. She can hardly understand how it is she is so little weary of her bed. I think it has very much helped her in this illness, that we have as much as possible kept the sick-room element out of it; letting her share all the out-of-door interests, and keeping up small amusements from hour to hour. The nurse's queer stories and her strange foreign ways of viewing things have all been a variety, and each day is full of small incidents connecting her with more active life. It is so different from an invalid wearying upstairs, however tender her nurse and friends may be, while the home-life goes on without her; here it is like a queer sort of picnic, a mixture of the pathetic and ludicrous, that is very strange. If we have callers in our salon, all the doors of all the rooms being open, she likes to listen; so altogether, you see, there is some good to be found even in what you might fancy were our privations. When C. and I are lunching on chicken and asparagus, with merry music coming through the blinds as the Bersaglieri tramp past, and Carolina comes up from her dinner laden with bouquets, and my father brings in fresh fruit and flowers and the latest 'Times,' and stays for a little gossip, and I prepare to cut the pages of a new magazine, while C. begins her embroidery, we often wish you could look in on us, and see how happy we are. My father spoils us with kindness. It is a great comfort to be blessed with a tolerable amount of cheerfulness; but really and truly, it is such a queer, odd, and pleasant life, now C. is better, that we cannot claim the least credit for being content; and when any great anxiety is lifted from us, life seems too full of thankful happiness. The mountains are a formidable barrier between us and home, with so great an invalid to be carried over them somehow; for we do not want a sea voyage, or wish to return by the Corniche; charming as that would be, it would be too much as though our journey had been a failure; and when once we are in Switzerland, all will be easy.

I hope you will recommend Carolina to all who may be in want of such a servant; she is admirable in every way—a most efficient nurse, and far better adapted to our wants than any English maid would have been; and as we were so fortunate in finding her, I am very glad we did not bring one from England. Her address is, *Madame Bernardi, 6 Via Benedetta.* She often travels as courier with ladies, when her charge is lower than as a *garde malade.*

XIX.

May 30.

It is late, and I want to go to sleep, as my painting-master comes to-morrow at half-past seven, but I must tell you a little of our charming drive to Fiesole. I am shut into my small chamber, a sort of three-cornered room behind C.'s, and with no outer door of its own, so that, when she settles to sleep, I am of necessity a prisoner till morning. My one window looks upon a terrace, and as that belongs to an adjoining salon, and is often occupied by other travellers, late or early, I can hardly keep it open for long together. The furniture is rather primitive; there is a colony of ants on the dressing-table, who have emigrated from the stone-flags of the terrace; our *black box*, a chest of drawers, a large bath, of which I have lately become the fortunate possessor, and the smallest of beds.

The mosquitoes, which are so troublesome in all our other rooms, have not yet found me out. They come in myriads from the Arno, though we carefully close our windows before lighting any candles. During the illuminations, I watched them in small clouds blowing round the lights before our rooms, and in our drives in the Cascine, we have seen them under the trees, or hovering over the fields, like a *thick mist* suddenly illuminated by a ray of sunshine.

As I have no table, I write on the bed, and if my letters are prosy, you must remember the difficulties under which they were accomplished, the thermometer at 78°, and mind and body sometimes rather the worse for wear after a busy day with callers, doctor, nurse, and my dear little patient, and walks or drives with my father.

There has been a little more air to-day to temper the heat, which has been so great every evening, that when sitting in C.'s rooms with the windows closed, one could not exist in anything but a dressing gown. Carolina laughs at me, and says :—

'Yes, yes; I knew how it would be; it is always so. My ladies, when they come to Italy, are talking of the lazy, dirty Italians, who never dress properly and don't wear stays. Wait a little, I think to myself, and by the end of a month they judge them more leniently'!

Sidney Smith's unfortunate hero, who wished to take off his flesh and sit in his bones, had a true understanding of the power of a southern sun. The hotel is so empty that they have given my father a large, cool, airy back room, and C. is moved into another next to her old one, which will make a pleasant change. Nurse carried her in her arms from one bed to the other. There was an amusing scene the other day, when some alteration had to be made in her mosquito curtains. One or two *facchini* and the waiter came into her room, with the carpenter, the erection being a rather mighty affair, as fortunately the beds in two of our rooms are large old-fashioned four-posters.

The men stood and looked at C., commenting on her appearance, much to her amusement. 'So that is the Signorina Inglese; she does not look so very ill after all, *povera figlia!*'

We drove this evening along the old road to Fiesole, and, close to his villa, met Mr. S., who was very kind, welcoming us at any hour to his house, and expressing much concern on hearing the cause of our detention.

We passed through the old town, a picturesque cluster of houses crowning the hill. The Duomo, a relic of the eleventh century, is a rude structure, with little of beauty or interest for the passing traveller. A roughly paved square, a few churches, the *Palazzo del Commune*, the episcopal seminary, and a straggling street of half-ruined houses, form all that remains of the old city of Fiesole.

The good modern road which winds up from Florence is a monument of the cleverness with which an Italian utilises what Providence has placed in his way. The road was urgently needed; but Florence was unwilling to assist, and as it was most important to the people of the hills that they should have easy access to the city below, the inhabitants of Fiesole, thrown on their own resources, fell back on the magical power of their *Libro d'Oro*. No one can be received at court who is not in some way ennobled; and as the fact of having your name inscribed in this wonderful book confers all such privileges, the demands were many from the bourgeoisie of Florence, and foreigners resident in Tuscany. Many a worthy and ambitious citizen went up to Fiesole and paid into the treasury his three or four hundred dollars, as the case might be, returning heated and exhausted as a *marchese* or *cavaliere*, but well content with his bargain.

We crossed the brow of the hill, ascending the one beyond till we gained a great elevation, and saw ranges of hill and mountain far away beyond Florence, which lay like a dim map at our feet. The country is barren and stony for Tuscany, but carefully planted in places with rows of cypresses, the road winding between them. On our right was the ridge of the Monte Ceceri, with its great sandstone quarries, *pietra serena*, which furnish the principal supplies of building materials to the Florentine workmen. There are curious ranges of caves hollowed in the rock, and the whole district must be well worth a visit.

An old castle, half ruined, towered above the dark green trees, and lower down we saw the solid masonry of what looked almost like a royal keep. It was formerly celebrated as the Castello Pucci, but had fallen into decay, when an Englishman, a Mr. L., purchased and restored it. He has carefully cultivated or rather planted the whole side of the hill with young trees; the ground was bright with golden broom, and bushes of pink and yellow honeysuckle with which we loaded the carriage. The air at that height was deliciously bracing, really almost cold, the sky was cloudless, with long bars of orange and crimson light on the horizon, the setting sun tinging the opposite hills with rose and violet, and just touching the dome of the cathedral on its way.

As we crossed the summit of the ridge behind the old chateau, the north wind blew keen and strong from the Valombrosan hills; I could only look longingly at them from a distance, as there is no possibility of our leaving C. for the many hours that would be necessary to visit them, even if we could enjoy such a pleasure without her. There is a railroad now, which, practically, very much shortens the distance, and the excursion can be made in a day from Florence; though to appreciate properly the beauties of the place, you should sleep in the convent or in the *dépendance* for ladies, and watch the sunset through the trees. It was very hard to give up all hope of seeing it during our stay in Italy; we had so often talked and dreamt of Valombrosa:

> .... Where, sublime,
> The mountains live in holy families,
>   And the slow pine-woods ever climb and climb
> Half up their breasts, just stagger as they seize
>   Some grey crag, drop back with it many a time,
> And struggle blindly down the precipice!
>   The Valombrosan brooks were strewn as thick
> That June day, knee-deep, with dead beechen leaves,
>   As Milton saw them, ere his heart grew sick
> And his eyes blind. I think the monks and beeves
>   Are all the same, too. Scarce they have changed the *wick*
> On good St. Gualbert's altar, which receives
>   The convent's pilgrims—and the pool in front
> (Wherein the hill-stream trout are cast, to wait
>   The beatific vision and the grunt
> Used at refectory) keeps its weedy state,
>   To baffle saintly abbots who would count
> The fish across their breviary, not 'bate
>   The measure of their steps. O waterfalls
> And forests! sound and silence! mountains bare,
>   That leap up peak by peak, and catch the palls
> Of purple and silver mist, to rend and share
>   With one another, at electric calls
> Of life in the sunbeams,—till we cannot dare
>   Fix your shapes, count your number! we must think
> Your beauty and your glory helped to fill
>   The cup of Milton's soul so to the brink,
> He never more was thirsty, when God's will
>   Had shattered to his sense the last chain-link
> By which he had drawn from Nature's visible
>   The fresh well-water. Satisfied by this,
> He sang of Adam's paradise and smiled,
>   Remembering Valombrosa. Therefore is
> The place divine to English man and child,
>   And pilgrims leave their soul here in a kiss.*

---

* Mrs. Browning—Casa Guidi windows.

I heard an amusing history of some very modern pilgrims, who were unfortunate enough to choose a *jour maigre* for their expedition. One of the party reminded the lady who was at their head of this circumstance, strongly advising her to postpone her visit; but to this she could not consent, as she was on the point of leaving Florence, and had no other day available for so distant an excursion; so she compromised matters by promising to provide a well-filled hamper, and they started, the younger travellers full of mischievous glee at the idea of bearding the lions in their den. As they approached the convent, a young monk came out to receive them with the usual courteous welcome, expressing much regret that, as it was a *giorno di digiuno*, they should be unable to show them the hospitality that they would have wished to display to their English guests. The lady hastened to assure him that they had foreseen and provided for the emergency, and had brought a good supply of poultry, tongues, and other viands with them from Florence, trusting to the good Fathers' providing the other concomitants of a meal for which the mountain air had given them no small appetite.

'Madame,' replied the monk, 'I extremely regret that the rules of our order will not allow of the introduction of meats, as you propose. We can neither furnish it, nor suffer it to be eaten under our roof.'

Here the impatience of a hungry Englishman gained the mastery, and one of the younger gentlemen broke in, unfortunately overlooking the fact that the priest was probably far more familiar with English than he was with Italian:—

'Never mind the old fellows; we'll have a jolly picnic in the woods, without troubling them; and eat our chickens, and send the bones, when we have picked them, to his *reverendissimo* the abbot.'

'Pardon me, Madame,' began the much enduring monk, 'but what the gentleman proposes would be an infringement of our rules which it would be impossible for us to permit. It grieves me deeply to say this; but I must beg Madame to promise that nothing of the kind shall be allowed. If you will venture to trust to our hospitality, we will do our best to serve you, and I hope you will have no reason to regret your visit to our convent.'

Of course they all acceded to the monk's proposal, and followed him to the refectory, where, in due time, an ample meal was spread. The cook excelled himself; and various were the dishes, all rich and excellent, though innocent of meat, which were set before them; and altogether the feast was voted so great a success, that not even one lingering thought was sent back to the carefully stored hamper, which was allowed to travel peacefully to Florence the next day, and served the party for an improvised supper at Lady ——'s lodgings.

While we rested the horses on the summit of the hill, I made a hasty outline sketch of the two castles and the valley below. Our coachman had never been to this spot before, and it was amusing to hear his expressions of delight, and his eager questioning of Signor A., assuring us that he should often bring travellers there, now he had learnt the road. Charmed to find a driver and horses so entirely *en rapport* with ourselves, my father gave him, in a weak moment, so large a *buono mano*, that I fear he raised the tariff in his own mind on the instant; while Signor A., who had often enjoyed this pleasant breezy solitude, looked half inclined to doubt his own wisdom in having found a guide intelligent enough to appreciate its charms, and ready to open a path to the general rush of travellers.

But there is no fear that the hills round Florence will ever be stormed; the Italians never walk and seldom ride, and they prefer the levels for driving, and the mass of travellers only go where Murray or the city crowds lead them.

Fiesole.

Above Fiesole.

Castello Pucci

A golden crescent moon was shining as we descended, passing carts drawn by noble dun-coloured oxen, and peasants returning from their work; and the city walls rose grim and dusk against the sky, as we drove under the great archway, and bumped and rattled home over the rough stones of the street. The bats were fluttering again round the old palaces, swinging through the air in their swallow-like flight above the great lamps on the Lung' Arno; and a row of boys on the parapet, sat cooling their legs in the breeze that blew from the river, while up and down paced slowly one continuous stream of people, smoking, talking, gazing at each other and the sunset, waiting for the moon, with busy fans fluttering, and smiling faces and little cries and laughter as a fresh carriage came dashing home from the Cascine, a waving mass of delicate lace and gauze, with a sudden scent of jessamine when the wind touched the great bouquets of flowers as it passed. Now and then a group of men would emerge from a side street, waking the echoes with a low musical chant, singing together as they sauntered homewards, or you heard the splash of the lazy oars as a little boat glided down the stream.

Often we have sat at our windows, flung wide open, leaning against the broad sill, and drinking in all the beauty and coolness, after the weary heat of the day. As the lights shone out in the houses opposite, and twinkled high up in the church towers, half among the stars, with long reflections quivering among the rippling water, it was pleasant to yield one's self up for a moment to the enchantment of the scene; the busy world below, the river in its calm flow, quiet and inevitable as life and fate; far above, midway between earth and heaven, the lights of San Miniato, where the dead are resting; and overhead, illimitable space, eternity with its beacon stars and love everlasting.

XX.

Florence.

I HEARD an amusing history, yesterday, of an English lady, a curiously strong-minded woman, who came to Florence many years ago. She was about forty years of age, and possessed some ten thousand pounds, which she invested in the purchase of a handsome villa and a considerable amount of land, which was, of course, of infinitely less value then than now. Despising *fattori*, and unwilling to trust her affairs to anyone but herself, she set to work in good earnest, farming her own estate, and selling the produce with great success; the Signora F.'s wine and oil fetching a high price in the market.

On her first arrival, and before she had decided on the purchase of land, she was at once marked down as fitting game, by the many keen and hungry fortune hunters hovering about Florence; and an Irish adventurer, anxious to retrieve his fortunes, distanced all other competitors in his eager attempts to secure her favour. But Miss F., being a wise woman, if not a very young one, was proof against all his attractions, and kept calmly on her way, looking out cautiously for a safer investment than any that matrimony, with joint interest, could offer, when matters were hurried to a crisis by the anxiety of the lover that so precious a prize should not escape him.

Determining that, if not to be won otherwise, she must be secured by a *coup de main*, he arranged his plans, and having found a willing confederate, resolved to carry her off.

Miss F. was known to be in the habit of walking alone in the outskirts of the city, and one morning a *vettura* was placed in waiting, with horses strong and safe to do any required distance in the shortest possible time, a driver already securely bought over, and the two men watching for their prey. As the unfortunate lady was quietly passing along the road, outside the Porta Romana, she was seized, and, spite of cries and struggles, gagged and placed securely in the carriage, which drove off at its utmost speed. But Miss F. was a woman of spirit, and by no means conquered yet; keeping her wits well about her, she struggled and fought her way to the window, freed herself from the gag, and alarmed the timorous driver by her outcries, till he, fearing unpleasant consequences to himself, yielded to her importunities so far as to draw up his horses, and offered no opposition when, breaking from her captors, she hurried with all speed towards the more frequented thoroughfares, where she could feel herself in safety.

She lodged a prompt complaint at the British Embassy, where she found but a cool reception, the young diplomatists looking upon the affair as rather a good joke than otherwise, seeing the heroine was neither young nor pretty enough to be interesting. A languid attaché received her in his shirt sleeves, and seemed so inclined to pooh-pooh the affair that she thought it wiser at once to lay the whole matter before the local tribunal. Here she met with better success, and her story was received with grave attention. The two men implicated were arrested, tried, and actually convicted of felony and condemned to the galleys, where they might have repented at leisure, if Lord Burghersh had not interceded for a commutation of the sentence, which was changed to perpetual banishment from Tuscany.

This brave-minded woman, meanwhile, after her little adventure, bought, as I have said, villa and farm land, and sat down under her own vine and fig tree, digging and planting, and raising crops with such success that her worldly goods were steadily on the increase. For the better protection of herself and her possessions, Miss F. became a naturalised Italian, and, to solemnise the act, agreed to be presented at Court. It was the custom for the Grand Duke to walk through the rooms, noticing those who had the honour of an audience. When he approached Miss F. he addressed her with much courtesy, ending his little speech by a polite enquiry as to her name. For once her presence of mind and consciousness of personal identity forsook her, and, unable to reply, in an agony of shy mortification she retired for ever from royal receptions!

The Duke and Duchess were in the habit of driving about in the neighbourhood of the city, and whenever they passed any villa or garden whose beauty especially struck them, they would ask permission to enter and visit the grounds, a condescension which made them very popular with the gratified owners.

Coming to Miss F.'s villa one day, they sent a servant to the house, and were gladly received by the lady, who, feeling secure and self-possessed on her own acres, forgot her old awe of royalty, and prepared herself to do the honours of her territory. Everything was examined and praised, till they came at length to the kitchen garden, when the Grand Duchess, with the easy tact with which well-bred people manage always to say the right thing at the required moment, expressed most gracefully her admiration of its order and completeness, bestowing especial marks of approval on a goodly row of red cabbages that were basking in the sun.

'Ah, Madame! what wonderful plants, and how much

finer than any that grow in the Boboli gardens; they are, indeed, astonishing!"

In a flutter of gratified pride, Miss F. begged her Royal Highness would accept some for the next pickling at the Pitti; so the royal footmen disposed of three enormous cabbages in the Serene Grand Ducal dickey, and they drove away, while the lady sent at once for her *arrocato*, and made her will, leaving all she possessed as a free-hearted offering to such discriminating royalty.

Shut up in her little domain, and busy with incessant work, she avoided society and ignored any religious observances; and on her death, some few years after, it was a matter of dispute as to which 'religion' could claim the right of reading prayers over her grave, while the poor lonely woman lay awaiting her burial; Catholics and Protestants hunted up evidence; the discovery of an English prayer book (the only work of devotion in the house) turned the scale, for want of any direct proof in the direction of the priests, and she was resigned to the care of the Church of England; her bequest to the Grand Duchess having been formally recognised, and the *legatario* put in possession.

Not long after, came a poor clergyman from England, claiming near relationship with the dead lady, with an old story of unhappy differences and estrangements, a brother and sister, 'a little more than kin and less than kind,' long years of separation ended at last, and bitter murmurs over the lost corn and oil and wine, that would have helped to fill his empty store-houses, and cheer his heart, saddened by much sorrow and many children and cares.

The case was represented to the Duchess, who at once declared that she would do nothing to defraud the Englishman of what he regarded as his right, but as she had entered into possession of the estate, and the villa pleased her, she was unwilling to resign it, but would commission her *procuratore* to appraise the whole, and the full value should be at once paid to the brother, who returned to his English home, we may well believe, with a heavy purse and a heart lightened in proportion. The gift of Miss F. to the Grand Duchess is still retained by her; a little green oasis for the poor exile when Florentine springs ran dry!

May 30.

Yesterday morning, I went with my father to Santa Croce, in the most frightful heat. A musical service was performed there for the repose of the souls of the Italian patriots who had fallen in the war of independence. It was a curious scene; church and army blended together to give all honour to the dead. The large building was crowded with people, who pressed against the cords stretching from end to end, and which formed a broad centre aisle, prepared as if for a procession. I do not know what the opening ceremonial may have been, but while we were there, no one ventured between the guarded lines but one or two ladies, dressed in deep mourning, each leaning on the arm of an officer, and carrying a small black bag, on which some white letters were worked, half filled with coins, which were shaken entreatingly as each fair petitioner solicited alms from the bystanders, the money being no doubt destined to pay for masses for the souls of the dead.

Soldiers with their band kept the great doorway; their music coming with a clash and burst of drum and trumpets into the monotonous chant of the priests, dying away as other voices took up the strain; the waves of sound floating up above the wreaths of incense and the crowds of people and the hundred lighted tapers to where, from the roof of the church, hung the great black banner with its white-lettered inscription: 'Blessed are those whose blood has been shed for their country.'

Carrying back one's thoughts to the time, still recent, when all the bloody and cruel oppression of a free-hearted people was an hourly reality; when the tidings of each day's struggle, each hour's failure or success, made our hearts throb with joy or sympathy; when, from the States of Italy, poor and down-trodden, all her strength and young manhood were pouring forth to dedicate themselves to her service,—one could well understand with what a tempest of proud sorrow and rejoicing the Florentines might have gathered round this altar already sacred to liberty, where a sacrifice of tears and blood appealed to the justice of heaven: but with time such commemorations lose their power; many of those who, six years ago, may have joined with their whole heart in the service, full of the fresh memories of shameless massacres, cruel imprisonments, ruin and death, from which nothing but God's blessing on their patient, brave endurance and noble struggles, could save them, now come to gaze, half with curiosity, half with a quiet content.

There was a constant stream of people going and coming through the open doors, all were quiet and orderly, but they looked as if they were merely assisting at a *spectacle*; there was wanting that electric chain of sympathy which can bind with subtilest links all hearts together; that pathos which, conquering all outward incongruities and absurdities, speaks straight from soul to soul, one touch of overmastering compassion which 'makes the whole world kin.'

In 1851, the tablets erected by the municipality, on which the names of the brave dead of Curtatone were inscribed, and placed by their orders on the high altar, were torn off by the command of Austrian Generals, who, when the relations and friends of the soldiers who had fallen, crowded into the church with votive offerings and garlands of *immortelles*, brutally ordered a wholesale butchery, their men firing upon the defenceless crowd, even within the walls of the sanctuary!

When the bloodless revolution of 1859 had been accomplished, the citizens once more gathered in the old church, which was hung with wreaths and banners, and many simple and touching inscriptions which would speak direct to the hearts of the people,—'The prayers which, for ten years, ye have offered to the Holy of Holies for your Italian brethren are already accepted, and they shall be fulfilled.' And 'To-day we celebrate the morrow of 1848!'

Mrs. Trollope describes, with eloquent feeling, this same celebration as she saw it in 1859, when Florentines were once more permitted, in Santa Croce, to do honour to the memory of those slain in the battle of Curtatone, where a handful of brave Italians kept at bay for hours the overwhelming force of the Austrians:—

'Behind the catafalco, facing the high-altar, stood a colossal statue of Italy, by Cambi; very admirable for its expressive power, and the earnest entreaty of the outstretched arms, imploringly lifting crowns of cypress and laurel up towards the Eternal Throne, and saying, in the words of an inscription, too long for me to quote entire: "Great God! who createdst me Queen of the Nations, who willedst in Thy justice that I, for my transgressions, should lie enslaved and trampled on; great God of Mercy! who now at length grantest me abundant measure of pity for my long sufferings, deign to accept the hallowed memories of these, the martyrs to their country!" And—as if in answer from above to the prayer of Italy, that the war now waged may "restore me to my own self again, link me in sisterhood with the nations, and lay the foundation-stone of holy universal Peace," so fervently expressed by the upward-gazing figure, with her crown of towers, and her noble, trustful, beseeching face—the great black banner above her head bore the words, "At last thou hast understood My ways. Thy hope was in Me only. With much love hast thou covered thy sins; thy

Santa Margarita.

The valley of the Emma, from Santa Margarita.

Lung' Arno.

faith has saved thee."' She adds, 'the whole pageant was a spontaneous heart-offering of every class of the people; and as such, it was invested with a higher beauty and a truer worthiness even than that which clothed it bodily.' I wonder whether the one in which we shared this morning was equally a work of the people; such offerings are so worthless when they cease to be from the heart.

After our morning in the church, we came back to the hotel to share our histories with C., and welcome some old friends from England, with whom we arranged a drive for the evening. We joined them, soon after five, at the Hôtel de la Ville, and went afterwards to Santa Margarita, the road being entirely new to them; visiting once more the curé's garden, and again returning laden with flowers from the curé's gardener.

We passed, as before, down a steep hill below the little church; our driver, like all Italian coachmen, having a great horror of a rapid descent, insisted on the gentlemen walking to lighten his carriage, and would gladly also have turned us out into the dust. It was amusing to watch the extreme care with which he guided his horses, all representations as to the desirableness of returning on our steps having failed. The shady trees by the roadside almost met overhead, and the high green hedges, bright with clematis, and honeysuckle, and ferns, looked like a bye-lane in Devonshire. The soil, too, just on this spot, was of a rich red, very pleasant to the eye; and there was nothing to remind us we were not in England, but a picturesque *contadino*, who appeared at a turn of the road, and the little shrine to the Madonna against which he was resting.

We drove the evening before in the Cascine, meeting more English friends returning from Naples, and having a good view of the king on horseback, riding about very simply amongst the people with only one or two attendants.

There are three or four fresh sets of Americans at the hotel; a gentleman from New York, with a pretty young wife, people we like; and an old man from the far west, with two clever daughters; they are all very fierce and bitter about poor Mr. Davis. It is not pleasant to hear ladies talk of their delight in the prospect of his speedy extermination; but one ought to make great allowances in passing judgment on the feelings or words of those who have suffered cruelly (whether by his means or not, time alone can decide), through many very dear to them, soldiers who were left lingering starved and ill-treated, dying of hunger within sight of the food sent to them by their friends. The English were terribly bloodthirsty about the Nana in the great Mutiny; and one of these American girls had had a brother in a Southern prison, who, after months of suffering, escaped indeed with life, but with health and strength terribly shattered—'Thanks to Mr. Davis'—and the pretty wife gives a little contented nod across the table, and says, 'Yes, I calculate we'll hang him!'

Wednesday, 31st.

I must finish my letter as quickly as possible. Another very fine day; but bright dry weather is so much a matter of course, I need hardly chronicle it. The mosquitoes were very troublesome last night, and my little room very hot, and I was glad to open the window at four this morning; I dare not do this in the evening, or these tormentors would find their way in. * * * The doctor has just been; he thinks C. much stronger. * * * I was dressed by half-past seven, and had an hour and a half's painting lesson. Signor Pompignoli speaks very good English, and if he gets stranded with technical explanations, floats himself off with French. We are just going to the Pitti for an hour; so good-bye!

XXI.

Florence, June 1.

Before beginning to journalise, I must thank everybody for to-day's budget of welcome letters; one from Aunt C. from B. I can fancy how disagreeable the city must have been during the races. How strange it is, that in England there always seems so much that is thoroughly bad mixed up with any of the amusements of the masses! I wish the people could learn to enjoy themselves rationally and simply, as they do here, where, in the midst of their greatest excitement, throughout the busiest Festa, we never heard an angry or coarse word, or laugh, or saw even one vulgar or drunken man or woman.

If you can fancy races and mock tournaments, and Bersaglieri trumpets in Arcadia, mixing in happily with the shepherds and shepherdesses, who all keep their crooks and their simplicity unspoiled, and drink their milk (with a little coffee in it, and an accompaniment of weak cigars), and make themselves very happy under the trees, with a bunch or two of ripe grapes squeezed into water, as an exhilarating fluid, you will realise the people of Florence enjoying a holiday in the year 1865!

The large cool rooms in the Pitti, which we visited again yesterday, were a delightful contrast to the square in front, where one would have been *cooked* in ten minutes. We drove there, walking slowly home through the shady streets; but it is really hardly safe now to go out except in the evening, or very early in the morning, which would not suit my father. I should very much like to make some sketching expeditions before breakfast, instead of always painting; but I cannot take Carolina out then, and it is impossible here to walk alone. Even quite poor girls are always accompanied. You see numbers of ladies on foot, shopping, marketing, &c., but never without a maid. When we first came here, C. and I went out several times together early in the day; but it would be considered strange for one to go alone.

I spent most of yesterday in C.'s room, reading to her, but the great heat seems to weaken the chest, which feels like an exhausted receiver, so that it is a considerable tax on one's powers; but now she is stronger, she can hold up a book and enjoy the variety of reading to herself, and we can get any amount of books from the library, besides our own 'Times' as a *pièce de résistance*.

At three o'clock, an honest working jeweller from a street near by, who had twice mended a chain for me without consenting to make any charge, came with stores of *pietra dura* work. We had not intended buying any, having such supplies of the same things already at home, but you know the old adage about 'idle hands,' and it was rather because I thought it would be an amusement to C. that I asked him to come.

She and I were alone, our valuable interpreter Carolina being away, and my father out, so we called him into the room, and spread out all the things on the bed, C. reviving fast over some charming little butterflies in brilliant coloured stones, which promised happily for Christmas presents in the future. The poor civil man looked ill and half-starved. C. begged I would explain to him that she was not suffering from fever or anything infectious. I am afraid my explanations were not strictly intelligible, as they ended with his saying:

'Ah, yes, rheumatism; povera Signorina! My wife

has had it badly these last three months, and, Santissima Madonna! how cruelly she suffers!'

We made some small purchases, asking him to come again, and settled once more to our books, when there came a knock at the door of our salon, which, now C. has been moved, is an anteroom to her chamber, so that she has the amusement of hearing what goes on. We have callers at all hours of the day, and as my father is out a good deal, I have to receive them, and can never properly enjoy a quiet spell of oil painting, because the moment I and my clothes are comfortably besmeared, there comes the little knock at the salon door. This time it was our American friend, the husband of the pretty Northerner, asking for my father, who had promised to help him in the arrangement of a Swiss tour. They had only three weeks to give to the mountains, and were honestly anxious to make the most of their time, and we were very glad to be able to help them. It was pleasant to meet with Americans who were willing to praise what they could appreciate and admire, not like F.'s acquaintance on Mont Vesuvius, who thought altogether 'it was a sad waste of valuable steam!'

As they were to start for Bologna in an hour, I sent Carolina to Vieussieux, and my father returned in time to give them a great deal of useful information.

We dined at the usual time, with Germans, Italians, and more Americans, a party of four, who had been wintering in Spain. We are far too apt to ridicule our neighbours, and to laugh at the peculiarities of Americans whom we meet for the moment at a table d'hôte, or on a steamboat, and with whom we perhaps only exchange a few words, amusing ourselves with criticisms on their bad English, forgetting that they do not profess to speak *English* at all. The Americans are a great nation, and there is nothing surprising in their having a language of their own. As well might a Dane or German smile over our ancestors' weak imitations of the old Teuton tongue. Thousands of their forefathers learnt their first baby utterances in English homes; but they have given the old language an individuality of its own, coining their own words (as many an Englishman has done, too, in our day); these words naturally partaking of the character and mode of thought of the people.

Trench says, 'slang is the growing end of a language;' what wonder, then, if the American language has grown more rapidly, and contains more slang than ours. It is the same, too, with the intonations and the very inflections and *sound* of the voice. No doubt they have various modifications, which enable the Americans themselves to decide who may pass as well-bred people; but it is difficult for an outsider to discover them. In England we have our own standard, most delicately graduated; and when a woman opens her lips, we know at once whether she is a lady, and in what class of educated people to place her.

It is absurd to condemn the Americans, simply because their voices are un-English, and their dialect natural to themselves. As well find fault with the deep gutturals of a German, the lively music of a Frenchwoman's words, or the soft syllables of the Italians. That which strikes us most in them is more the peculiar intonations of the language as a whole, than the use of the provincialisms, of which so many of the vulgar English are guilty.

I like strong individualities in nations, and that people should be willing honestly to confess and to respect them. It took us a long time to learn that it was not part of an Englishman's creed to despise a Frenchman because he was supposed to call 'frogs' *crapauds*, and even went so far as to eat them; while John Bull felt a frog was a frog, and no one but fools would try to make out otherwise!

I am only writing of well-educated Americans, as those we meet are for the most part. When 'Shoddy' has invaded Europe, as it threatens to do (and people tell us all the steamers are engaged already for the autumn, by hundreds and thousands of these great capitalists), it can only be expected to hold its own with the marvellous specimens of cockney and Manchester English who are also making a tour on the Continent.

The Americans look at things a good deal *en masse*, and rather from a utilitarian point of view; like that celebrated Yankee who, after a careful study of Rome, pronounced it to be 'a fine city, but a good deal out of repair!' They have a wonderful power of generalising their impressions, take the world calmly, and are never to be caught at a disadvantage, or to be made honestly astonished at any marvel of nature or art. The latter, as a nation, they cannot understand; and it is curious to hear an educated and intelligent gentleman describe to you his first impressions of such a gallery as the Pitti, or his morning's visit to a studio, and the principle on which he has selected pictures for purchase. As a rule they prefer sculpture, because, as my father declares, 'they calculate they can walk round it!' but the statues must be sufficiently draped. Gibson had to put a small garment on his little 'Sleeping Boy' before it was available for the American market!

In the evening we drove to the San Salvi, formerly a convent, an unpretending old building without the walls. In the refectory is the beautiful *Cenacolo* fresco by Andrea del Sarto. We have now seen the three finest in the world; the Leonardo da Vinci at Milan, the Raphael in the Egyptian Museum, and this, which is far too little known to tourists. Murray does not mention it, and hundreds pass through Florence without visiting it; and yet it is only a ten minutes' drive beyond the gates. The painting is very wonderful, the colours are still fresh, and the bold outlines well preserved. Photographs of it are sold at the convent, and may also be obtained through Goodban.

Though the arrangement of the whole is similar in all three, there is a good deal of variety in the treatment of the figures; Christ clasping the hand of the St. John, and giving the 'sop' to a figure on his right, which I think must be intended for Judas; but, if so, there is none of the distortion of crime usually given to the face; it is more the unhappy disciple before Satan entered into him. Andrea del Sarto's paintings have a great charm, in their power of drawing, and the harmony of the whole. The colouring, though often very rich, is not so uniformly brilliant as in the masters of the Bolognese school; but there is a truthful effect of light and atmosphere, each colour being necessarily toned down to keep it secondary to the perfection of the whole; so that, what you lose in detail, you gain in general effect. We drove round the town, through rather dusty lanes, that would be all the pleasanter for a good shower, passing the Cemetery, which we visited again, to gather some flowers from ———'s tomb for his mother. We found Mrs. Browning's grave, which we had missed in our former visit; it is covered with a broad low slab of marble, probably designed to serve as the pedestal to a more finished tomb.

We had another beautiful sunset; it was very lovely and peaceful in the Cemetery; with the dark cypresses and the bright flowers, the glorious figure of Hope rising over a grave, its radiant face turned towards the sky, and a flush of pink and violet light bringing the earthly hills and the heavens together in one glow of evening beauty.

The birds were singing, and the wife of the *custode* was sitting near the gate, with two pretty dark-eyed children lying against her knee, tired of play, and one or two solitary mourners were moving quietly about, tending the roses or laying some little wreaths of *immortelles* on a tomb.

We returned from our last night's drive about eight o'clock, in time for C. and me to have our evening meal together. We are still faithful to our cakes and milk, and eat an amount of fruit that astonishes the Italian servants, the Florentines partaking very sparingly of the same; and no wonder, when the rest of their diet is salad and cabbage soup, vinegar and sour wine, any extra 'vegetable acid' being decidedly unnecessary. Our daily supply includes English and little wild strawberries, large white and red cherries, and small apricots, neither very good nor very wholesome, and figs, which are just beginning to ripen. C. has been indulging in a private purchase of cream, through Carolina and the milkman, of whom any quantity may be ordered. A small allowance was served to us in a little flask, about three inches high, at the cost of a few pence, with a primitive cork or stopper in the shape of a screw of paper, placed in the tube.

The doctor has discontinued his evening visits, which looks like progress, and we hope now to get away before any very great increase of heat, though the possibility of any journey over the mountains looks very far off.

We hear constant stories of the outrages perpetrated by the brigands of southern Italy; all our friends returning from Naples whom we have met here being full of excitement over the perils they have successfully braved, or the misadventures of their less fortunate acquaintances. Dr. W. told us a good story the other day, which has the additional interest of being a true one.

An Englishman engaged in the superintendence of some mines belonging to the late Duke of Parma, was informed one morning by his servant that a man wished to see him, who declined mentioning his business to anyone but himself. Being much engaged at the moment, he felt half unwilling to admit him, but at last concluded to do so. The Italian stated that he was in possession of information that would probably save him from being robbed, if not murdered, but that, before disclosing it, he required to be paid a certain sum, which he named. This proposition was at once declined, but Mr. M. told him he would give him some trifling gratuity, and, if in the end his suspicions turned out to be well founded, and his story proved to be correct, he should be rewarded liberally.

To this the man assented, adding, 'You are an Englishman, and I can trust your word; and now listen, and judge whether I speak truly. You propose to visit the mines at ⸺, on the eighth of this month. You have engaged Beppo Quattrini's vettura and horses, and he himself is to drive you, and you will carry with you so many hundred scudi, for payment of wages now in arrears.

'It is true,' replied Mr. M., 'that I am going to the mines on the day you name, and also that I have engaged Beppo Quattrini and his carriage, and that I may possibly have some money with me, though what interest all this may have for you I cannot imagine.'

'Listen, Signor! On reaching a stone bridge, some distance from Parma three men will attack the carriage, and Beppo *will offer no resistance.* You will be robbed, and probably murdered, for dead men tell no tales. And now, does the Signor think my story worthy of credit?'

'Not in the least;' the Signor was not to be moved by such a *papera*; with an Englishman's cool disbelief in danger, and dislike of being forced into a ludicrous position, and swindled by an adventurer working on his fears through a clever story, he pooh-pooh'd the whole affair, and dismissed his informant. That night he dined with the Grand Duke, and mentioned casually his visitor of the morning. The Prefect of Police was at the table, and when dinner was over, he drew him aside, and assured him that such a warning was not to be lightly disregarded.

'You may be satisfied every word the rascal said was true; but you need not trouble yourself about the matter. Continue your preparations as before; keep a silent tongue in your head; and leave everything to me.'

The morning arrived, and Beppo drew up his horses at Mr. M's., where the Prefect had previously called and given his instructions.

'You will take a gens d'arme, whom I shall send you, as your valet, and he will ride beside the coachman; at the nearest village two more will be in waiting, who will enter the carriage, and accompany you to the mines.'

The start was made. Beppo dared not object to the companionship of the servant, who, as they approached the village indicated, drew a pistol, and placed it to his ear.

'Listen, *briccone!* you are sold; you are a dead man! if you wink so much as an eyelid, I will blow your brains out, *maledetto!*'

As they approached the fatal bridge, three men, armed to the teeth, rushed from their ambush; one of them seized the horses, but was fired at by the man on the box, and mortally wounded, though at the moment he succeeded in making good his escape. The two other miscreants made a dash at the doors of the carriage, and were instantly shot dead by the men inside. The coachman got off in the end more easily than he deserved; and it is to be hoped, for the good credit of Englishmen, that Mr. M. discovered and rewarded the man who had given the warning, and undoubtedly saved him from certain death.

This gentleman seems to have borne a charmed life, and many are the stories told of his wonderful adventures and escapes.

I must not end my letter with this tragical story, but

tell you something of the pleasant parties that are the fashion amongst the Florentine English. People gather in the grounds of one of the beautiful villas round the city, about six o'clock—gentlemen and ladies in morning dress—and play croquet, till the increasing darkness sends them into the house for coffee and music and pleasant chat. We hope to go to one of these 'receptions' next Wednesday.

There is more air to-day, and it must have rained in the mountains, as there is fresh sand in the Arno. C. has been allowed to sit up for an hour, and she enjoyed the change, and worked and looked out of window, propped up in an easy-chair, while I read to her. My father and I drove to the Spezziaria of Santa Maria Novella, where I bought some more orange-flower water. As I write, the room is filled with the scent of orange-blossoms and magnolia, sending me nearly to sleep.

We saw the distilleries, and a heap of pink rose-leaves on the floor, and three monks busy stopping up ends of bottles with corks and wax, and coloured papers, pleasantly at work; and through all the beautiful rooms, there was a subdued rich smell of unknown essences, with gilding, and pictures, rare old paintings, bright mirrors, and draperies, and a pretty little court with statues and flowers, and through a window one looked out on a dim, quiet cloister, where visitors are not allowed to enter.

We drove to the Piazza dell' Independenza to call on the English clergyman, who has been most kind in his offers of help, and then to the Piazza Santa Croce, to see Signor Pompignoli. I was delighted at having succeeded in inducing my father to visit the studio, and very much hoped he would be persuaded to purchase a copy of his favourite *Gran Duca Madonna*, but the only one we saw was unfinished, the work of a pupil, and it had not sufficient beauty to tempt him into giving an order for it; in which case, Signor P. would have worked at it during the summer, finishing it himself from the original. I am sorely tempted to buy a copy of the *Cardellino*. There is one very good one in the Tribune—the artist being still at work on it in the Uffizi, but he asks one thousand francs, which is more than I care to give.

The price of these paintings has risen wonderfully in the last few years; the work is badly paid as it is, and the poor artists toil day after day, using up every available moment during the hours when the great galleries are open. We looked at a beautiful copy of the *Seggiola*, which the Signor has just completed, and which was at once bought by a Scotch gentleman. We congratulated him on the speedy sale, when his eyes filled with tears, as he looked lovingly at the fair-faced Madonna—saying, in a sad voice—'It is hard to me to part with it; I am growing old, and shall soon need rest; I shall never paint another at the Pitti; my turn does not come again for ten long years!'

It is easy to obtain the necessary permission to copy the originals in the great galleries; but a weary time of waiting is often entailed on the artist before his turn comes round. There are many ways of obviating this however; the artist, whose right it is to paint a Perugino or a Raphael, may permit another *pittore* to place his easel in position, and work at the same time. Also, during the early days of autumn, when the galleries are closed for cleaning or repairs, a few are always able to gain admission and accomplish their work, by relinquishing the delights and repose of *villeggiatura*.

Some of the picture-dealers seem unable to afford the luxury of a conscience, and unwary travellers are too often made their dupes. We heard a story of a gentleman who was anxious to purchase well-authenticated originals, and who called at a studio where he had been told he should find the class of painting he required. After looking through a number of rather worthless pictures, the Englishman demanded whether the dealer had nothing better to show? and was then, with some amount of mystery, assured that he did indeed possess a great treasure—(say a veritable Giorgione, I forget the name of the particular great master,) and the Signor should at once see it. The Englishman was a real connoisseur, and not easily to be imposed on. He satisfied himself that the painting was a thoroughly genuine specimen of the artist whose name it bore, and after making the purchase, he determined to provide against any attempt at deceit, by writing his name, and placing his own seal at the back of the canvas. Unfortunately, he omitted to take it away with him then and there; it was left in the hands of the dealer, who was at once to pack it for carriage.

On his return to England, he opened the case, eager to enjoy his treasure, and exhibit it to his friends; but there was something wrong about it. It could not have suffered in the journey; but somehow the colouring was less soft; the whole effect had ceased to please him; the charm of the picture was gone. He examined the canvas; there were his name and seal as he had placed them, unchanged, and untampered with; he was fairly puzzled. It is very difficult to carry two paintings in your mind, an original, and a *replica*; and more difficult still to remember your *first* impressions of the one you have still before you; but he felt sure there was some very strange and startling difference between his Giorgione—there before him, and the Giorgione he had purchased in Florence. Determining to set the matter at rest, he wrote to a friend residing in Tuscany, and requested him to endeavour to hunt out the truth. This was soon discovered, and proved anything but satisfactory to the unhappy victim of a most successful swindle.

His friend called on the picture-dealer, and requested to see any particular treasures he might have to dispose of. The little farce was gone through, as before; a host of worthless copies were exhibited, and then, when his customer still made anxious enquiries as to any originals he might be possessed of, an authentic and beautiful Giorgione was drawn forth from its hiding-place, the *stock*, and veritable *original* of the establishment, and the exact *double* of the painting in London. By a clever trick, the real Giorgione was fastened over a copy, on the back of which the unfortunate purchaser had written his name, which signature had been duly forwarded to him!

XXII.

June 3.

I FEEL as though my letters must be just alike, almost ludicrously so, as the days are so much a repetition of each other; a painting lesson, many hours in a sick room, a dull table d'hôte dinner, and a drive. I am weary of describing Florence from the distance, or from the hills, or the street life around us, and the wonderful sunsets, which must always be painted in the same words, as I have no talent for inventing fresh adjectives. When we could both write, there was some chance of a little freshness in the letters; but now I have a humiliating sense that you will weary of my endless monochrome days, and 'purple and orange' evenings. I must tell you to-day, for a little variety, that the clouds are blowing up for rain, so I shall be able to indulge you with a run upon greys, as a change.

Yesterday, after finishing my home epistles, while my father was busy writing to F. in the Tyrol, I went out with Carolina for some necessary shopping, which it is amusing to get through with her, as she always gives the people a third less than they ask. I wanted to buy a common paper fan for C.'s use, the real value of which was about sixpence. The *bottegaio* demands two francs; whereupon Carolina lays it down with an impatient shrug, and a 'Via! via! it is not worth a brass farthing; of what is the man thinking! Leave it, Mademoiselle, and let us try elsewhere.'

Then the salesman, who has no intention of letting us go, says:

'What will you give, then? it is well worth the two francs, as the Signorina may see.'

Nurse offers forty centesimi; the man comes down to a franc and a half,—eighty centesimi,—she is not to be moved. At last, with a cheerful air, as if it were quite a new idea that had suddenly occurred to him, he proposes seventy centesimi, and she gives me a little nod, as much as to say, 'You may take it now, that will do very well, it is about what it is worth;' and the shopkeeper folds up my purchase, and wishes us *buon' giorno*, and we part quite excellent friends. He has not the least the air of a man who has been beaten, as he knew from the first he should never get what he asked; and we walk home without the disgraceful sense of having cheapened an article unfairly, and cheated an honest tradesman. Carolina was buying a pair of boots for herself the other day; she was asked nine francs, and quietly laid six on the counter, which were received with a satisfied '*grazie*,' as, with the boots in her hand, she left the shop.

We walked a great distance to find some particular kind of cake, which I thought C. would like, and came back laden with parcels, books from the library, and a bunch of bright flowers, given me by one of the flower-girls on my way.

These women supply Florentines or travellers daily with the freshest and sweetest of flowers, pressing them upon you, regardless of payment, and expecting in return a present in money when you leave the city. My father

Elise.

Florentines.

would never allow us to accept them without giving at once a few centesimi in payment; and often, when we had chosen all we wished to buy, we had to toss back the masses of roses and jessamine and sweet clove pinks that had been pressed upon us. Once, as we were driving rapidly through the city, a large bouquet was flung by one of the flower-girls into the carriage, without our even being able to recognise her.

Yesterday, at six, we drove out, passing by the Porta San Gallo, and winding in and out among the villas which surround the city, driving towards the Fiesole hills, without ascending them, and trying to find new roads and windings among the bright trees and gardens. We passed one very large and handsome house, an old Medicean villa, its soft, warm, rich brown colour and quaint style of architecture making it very picturesque, spite of grand new grey stone lodges and pillars at the gates. There is a fresco on the walls, painted by Mr. Watts, when Lord Holland lived here, to commemorate the fate of an unhappy doctor, who, summoned in haste to one of the old Medici, and failing to cure him, was seized by the enraged and grief-stricken servants, who, believing that it was in a leech's power to kill or cure, and seeing that this particular medico had not cured their master, but rather stood by calmly while he died, threw him down a deep well in the courtyard, where he perished miserably.

It is a curiously-built house, with a covered gallery full of openings, like windows, under the roof, formed by a second outer wall which ends in machicolations where it rests against the main building. It belongs, now, to an Englishman, a Mr. S., who owns a great deal of the land around it, and derives a large income from copper-mines in the neighbourhood. He has been a munificent patron to modern Florence, having been the principal contributor to the expenses attending the erection of the new front to Santa Croce, and he has, I believe, offered to assist largely in a similar work for the Duomo.   *   *   *

XXIII.

Florence.

I MUST tell you something of Nurse's story, as I have heard it from herself, or in her talks to C. She is an Alsatian, a clever, cheerful-spirited woman, with what the Americans call 'faculty;' a capital companion, very amusing and original, full of resource, and of clever little plannings to meet every emergency. She tells us a great deal of the life of the people, and their ways and habits of thought; and is never dull or prosy. She is about the middle height, upright and strong, and must have been handsome in her youth, but her face is slightly marked, as though tanned by a hot sun. We asked her once how she had injured her complexion.

'Why, you see there were ten or a dozen of us children, and I was the youngest, and one day my mother went out, and left me to the care of an elder girl, and she, wearying of her charge and wanting to play, ran out with the others, taking me with her; but she was soon tired of me, I was so small and could not run, and so she threw me into a lime-kiln in passing, and, *ma foi!* I should soon have been finished then, if a man had not heard my cries and picked me out; but I was burnt here on my neck, as you see.

'We had great misfortunes, when we were little; my father had been on a journey, and was returning home, when he stopped to rest at a way-side inn. It was growing late, and he was still many miles from the village where he lived, and in an unhappy hour he yielded to the landlord's persuasions to remain with them for the night. They put him into a room where a girl had lain who had just died from typhus fever, and my poor father caught it, and came back to us only to sicken and die. The dreadful disease fastened upon our house, and one after another the children were struck down, and at the end of eight weary months of suffering and death, our mother died, too, fairly worn out by her sorrows, leaving me and one or two brothers and sisters to struggle on as we could, or be buried away like the others.'

Somehow, Carolina, who, as the lime-kiln story shows, must have had some strong vitality in her, did manage to live, spite of all difficulties, and grew up a handsome-faced girl, and coming south as maid to Lady Holland, married an Italian courier, and lived with him in Florence, where their four children were born. The husband was kind and good to her, and they were happy, though not over prosperous; but his health failed, and after a year of anxious watching, he died, and she had to face life again, alone.

She hired a floor in a quiet street, made a home for her children, and worked for them night and day, sometimes as a *garde-malade*, sometimes as maid; living with one lady for many years, and going away with her to Ireland. This was when the children were growing up, and she was hoping the sons would be able in their turn to work for her; but the eldest married early; and a year ago, when she returned to Florence, after travelling with an invalid to Paris, she found, to her dismay, that her youngest son also had set up a home for himself.

The laws are very strict as to marriage, which is impossible without the consent of the parents, and poor nurse had imagined that there was safety in absence.

'It was hard on me, Mademoiselle. These young things, they forget how others toil for them; there is very little gratitude in the world. I had worked for them all these years, and thought of them before everything; and I might have married too, often and often, Mademoiselle. There was old Luigi, he had a thousand francs income; it is true, he was lame, but he was amiable as an infant, and I should have lived like a lady. And when I was in Ireland, there was a farmer, he wanted me badly enough, and I should'nt have minded his red hair, and I am clever with cows; but there were my children, and Carlo had said to me when he was dying: "Carolina, marry again, if you will, but promise me never to leave the children." And I never will, Mademoiselle; and I never shall take another husband, never; not but what I could find a good one ready to-morrow, but Carlo and I were very happy, and I loved him.

'Well, when I came back from Paris, there was my boy, and he only a boy of eighteen, married, forsooth! and my daughter-in-law old enough to be his mother, and proud enough of my big son. I spoke my mind to her, I can tell you, Mademoiselle; a woman like that, who ought to have known better, marrying my boy, deceiving and cajoling him while I was out of the way. "Are you not ashamed of yourself to marry a lad like him, and live on his earnings,"—I said to her—"he that ought to be helping to support his sisters, and let his mother rest in her old age?"

'Not that I am so very old, Mademoiselle! I have plenty of work in me yet; but it was just as well to say it, and she'd past her youth, too, by a long bit, and knew it. And how do you think they managed the marriage? They got their papers all right from the priest, and then, when they were questioned, he said his father was dead, and his mother had run away!

'That was a little cruel, I think; but they forget easily, these children. And so the law looked on him as an orphan, and they had their own way.

'Well, that was last year, and now the conscription is coming, and he is sure to have to go, for he is so tall and strong and healthy, he has'nt a chance; and his wife comes crying to me, with a piteous story, and I say,

"Very likely, indeed! that I could raise the money; and where am I to get it from? You should have thought of this a little sooner when you married a great child like that, that hadn't served his time, and his mother away and never knew it."

'The conscription falls hard on the poor now, and you have to give ever so much for a substitute, 3,000 francs; and in the old days you could buy a man off for 500; and the prices of everything are raised, and it is difficult to get along at all with your head above water.

'Well, they must manage for themselves; I want all my savings for my daughter. She's a good, quiet girl, a dressmaker, and these seven years young Becchini has been making love to her. He is well off, and his father is a master printer, a worthy man, and he would give his consent to the match, but the mother! It is she who is the grey mare! She comes from Arezzo, and brought some money with her, and that fortune is the bane of the house. Her eldest son married three years ago, and she never forgave him, and has never set eyes on his wife and child; and now Joseph is her darling, and he is to have the money if he will only give up my poor girl, and choose some one grand enough for his mother. It is not that he cares for the money, and he is very fond of Elise; but he is a good, easy-going young fellow, who loves his mother, and does'nt want to cross her, and would do anything for a quiet life. And so they go on, year after year; and I say that is all very well, but I want to marry my daughter. She's young and handsome, and she and her sister are all alone when I am out earning their bread; and it is time she had a husband to take care of her. "Marry her, or leave her," I said to him; "she's wearing herself out with the fretting and anxiety."

'I told you, Mademoiselle, that Becchini is a printer; and not long ago, the priests wanted offers for a contract for some of their publications, and strangers came to the city from Bologna and Milan; and, one evening, as he was going home, and it was growing dusk, one of these *scélérats*, who was angry because he had agreed to do the work for a lower sum, struck him in the side with a stiletto.

'He came to us in a carriage, white as a ghost; and we cut off his coat, and dressed the wound; and my girl said nothing, but just a day or two after, when he was better: "Mother, if anything happens to Joseph, I shall die." He's like the sun to her, Mademoiselle; and she would just pine out of life, if I tried to part them. I knew that before; but I saw it plainly then, and made up my mind they should be married before another year was over their heads.'

May 27.

This morning Elise called to let nurse know how affairs were progressing. The bridegroom dares not tell his mother the whole truth; but she begins to see it is inevitable. The marriage will be very soon; but, even then, Joseph thinks it may be better for his wife to live at home, as before; that this formidable mother-in-law need not be told at once that the affair is arranged for good and all. If the *day* can only be kept a secret from her, there will be a few minutes' more breathing time for her family. Poor things! this Arezzo heiress is a terrible despot.

Elise is busy with her preparations; it is customary for a wife to bring bed and linen for her room, the husband providing the rest of the household goods; and poor nurse is busy collecting money for these serious expenses. The girl will have forty francs dowry from her parish, her priest giving the needful certificate of due attendance at mass and confession; then an old lover of Carolina's, who has lapsed with time into a family friend,

Carolina Bernardi.

Waiting for the Bride.

adds forty francs; but there is a black silk dress to be purchased, the great ambition of a young Florentine. This is for the wedding, being considered very full dress. It is curious how much it is worn everywhere abroad for fêtes and marriages. I remember when Fraülein W. came to England, she showed us a gorgeous silk dress of some wonderful tartan, which she said had been especially purchased to wear at church, as she had been told the English considered black dresses *triste*, and always wore the gayest possible colours on Sunday! My father promises the wedding bonnet.

May 30.

Elise has come to thank the Signore for his present. She has a beautiful olive-coloured face, and wonderful eyes, and that calm, composed manner that gives such dignity to these Florentine girls. The marriage is fixed for next Saturday; it is to be a very quiet one, on old Madame Becchini's account, and because, nurse says, 'these young men are so foolish, they will not go to the church in broad daylight; they are too shy, the great geese! and besides, Mademoiselle, you see they do not care for the priests now, or the Church's blessing. It was different when I was young. Carlo and I were married in the morning, at mass; and that brings good luck, they say. I am very angry with Elise for giving in to his notions. But, after all, what does it matter? My husband died; so it did not do much for us!'

And so the wedding is to be in a little chapel of their parish church, about eight o'clock; but they make a mystery even of the hour, and hardly anyone will be present, and the bride will not even wear her new bonnet. They go dressed just as usual, and get married as quickly as possible, and then, the next morning, Elise will wear her new clothes for the Sunday festa, and she is to come here to see us. We are quite excited over the romance of the whole affair, and I can see Carolina is very nervous. It will be dreadful if the cruel mother-in-law should appear at the last moment—perhaps even part the lovers at the altar! What a pathetic possibility! I must keep my letter to tell you the end of the story.

June 4.

Last evening I walked with my father to the church of Ognissanti, which is very near the hotel, and discovered the little side chapel where the marriage was to take place. A small, dirty, stuffy, little hole, with a bench or two against the walls, and an altar just dimly visible in the light from the doorway. We explained to a man guarding the entrance that we were there by invitation, and he fetched a long dip candle, and lit another on the dusty old altar. This chapel is the place where dead bodies are brought before burial, and altogether it looked very weird for a wedding!

We were there soon after eight. There was a fusty funereal atmosphere about the place, like an ecclesiastical morgue, that made us both prefer waiting at the door, which we did, anxiously looking up and down the street, and fearing things were going wrong. The passers-by often turned to watch us, fancying, probably, that we were hopelessly waiting for a bridegroom! My white dress, which was, of course, only an ordinary costume, was certainly the only thing like a wedding.

At half-past eight, a fat old Padre, in brown gown and cowl, and roped waist, and sandals, took his place in the chapel, and patiently dozed in a corner, assuring us from time to time, that it was '*molto troppo tardo*,' and more than Job himself, to say nothing of a Franciscan, could be expected to put up with. So my father thought when nine o'clock rang out from the churches; we paced up and down the street, hoping each distant carriage was bringing them nearer, or eagerly scanning the approaching groups; but at last, wearied out, we turned homewards, with sorrowful thoughts of our poor, pretty bride, and fearing everything from such unaccountable delay; while the brown monk and the man with the candle waited on, and the holy tallow on the altar (consuming all this time at a fearful rate, to be paid for out of the fees) guttered down the little taper as if in feeble expostulation of such useless waste of time.

C. and I had our milk and fruit together as usual, and I read to her till past ten, when suddenly Carolina appeared, rather excited, but very happy and thankful to have all well over, and the wedding safely accomplished at last. Joseph's brother had kept them waiting, having driven into the country that morning to fetch his two years' old child to spend Sunday with its mother; the little thing, after the usual Florentine fashion, being out at nurse in a village. The brother was wanted as one of the witnesses, and as he was an hour behind time, they did not reach the chapel till half-past nine, and then, she says, it was all over in a minute. The old Padre read something in Latin, which meant, 'Wilt thou take this woman,' &c., and young Becchini, the printer, said 'Yes, yes,' as though he considered that might have been taken for granted, and the monk read their papers and blessed the rings, and was paid ten francs for his trouble and the tallow candle, and the bride cried a little at this ending of the seven years of doubt and love-making, and kissed her mother, and they all went to nurse's house, and drank a little Marsala, and ate some English biscuits by way of celebrating the event, and then her two girls and the bridegroom brought her back to our hotel; Signore and Signora Joseph Becchini taking up their abode in the house in the Via Benedetta.

Carolina had found them at dinner, her daughters and the lover, busy over some roast beef and beans, when she joined them at her home in the evening, fortifying themselves for their expedition to the church.

We gave her a glad welcome. She said 'It's well over, thank God for it,' and prepared to settle C., but her hand shook a little, spite of her attempts at hard-heartedness, which she has tried to keep up all the time, and she upset the iodine, and looked half inclined to cry over it, saying, with a half sob, we must forgive her, she did not know what could be the matter with her; but Elise had been a good girl, and she was very thankful, and very happy about her, and then she fairly cried; and we sympathised with her, and talked it all quietly over, to our mutual content. Was there ever a woman yet who did not enjoy a love-story—but this is the end of ours. Good night.

XXIV.

June 6.

E. is busy with her painting, and has asked me to write to you, as she cannot undertake to describe a second illumination. The festa of San Giovanni was brought forward a little this year, and combined with the rejoicings over the *Statuto*. The great day began by an early review of all the troops in the Cascine, by the king. I lay in bed and listened to the tramp of the men as they passed under our windows, and was finally taken up in nurse's arms and carried to an easy chair, just in time to see the last of some very dusty soldiers in blue. The Bersaglieri got mixed up somehow with the cavalry, and were shouted at and made to run backwards and forwards, like sailors vainly hunting for a mast up which to swarm, as there seemed no possible egress for them on terra firma, fresh troops rapidly closing up every avenue behind.

The Ponte alla Carraia was covered with preparations for the fireworks, and all along both sides of the Arno up to the Ponte Santa Trinità were lamps for the illumination. As the day wore on crowds of respectable-looking, well-dressed people passed and repassed on foot and in carriages, and the river was alive with boats of all sorts and sizes, filled with men and boys, who delighted in rocking them backwards and forwards, at the imminent risk of an overset, their shouts of song and laughter coming to us joyously over the water. The fireworks at nine o'clock, which I was allowed to sit up for, were very grand, flights of rockets and fire balloons following each other in such quick succession that the whole sky seemed studded with new stars, and then a sudden shower of red, white, and green lights, the whole length of the Arno, even the bridges being marked out for the moment with a beading of light, until the smoke made everything dim and weird looking, with globes of colour flying about like illuminated cannon balls; and then came a succession of most startling reports as each separate point of light exploded, making us feel as if the hotel were blown up and we in it, as we watched the startled crowd below falling back half suffocated on each other.

Our bride and her little sister have just been here, she looking very bright and handsome in her wedding finery, which she would not wear last night. Fanny was not allowed to be at the wedding. The Florentines do not consider bridesmaids a necessary part of the marriage ceremony, and it is not *comme il faut* for young girls to be present.

The king passed our hotel in the evening, in state, with his outriders in scarlet. He was very well received, the people clapping him as he drove by. He had had to sit solemnly in the Piazza Santa Maria Novella, and witness the chariot races which take place each year in honour of San Giovanni, to the great delight of the good citizens of Florence, who crowd at their draped and decorated windows, and fill the raised seats erected round the square, staring at grand duke or king who come thus to pay their respects to St. John or the people.

When the fireworks were over, the lamps were lighted, and the sight was a very lovely one. All the old buildings and bridges were brilliantly illuminated, lanterns and torches moved along the river's banks, and on the water, boats with bright-coloured lights and festoons of flowers forming arcades above them, glided up and down, the oars keeping time to the voices in their slow soft chant or merry song far into the night.

E. drove with Carolina through the city, to see the Piazza della Signoria, with its grand, rugged old Palazzo Vecchio tower shining far up in the heavens, in a soft veil of light; the pillars of the Loggia de' Lanzi, wreathed with lamps, and an illuminated verandah in front of the great post-office, all a-blaze with the Italian colours. A low hum of gently modulated voices rose and fell with the movements of the crowd, or was drowned by a fresh burst of music from the bands, which were stationed in almost every square of the city. The great Palazzo della Communità was all a-flame with lamps and brilliant transparencies and groups of banners, and there was rejoicing shining out everywhere, except where, in the long streets, the stream of light was broken by a dark mass, the home of some staunch *Codini*, who scorned to make merry with the people over the freedom they begrudged them, and closed their eyes, and hearts, and windows to the gladness of grateful Florence.

I am really better, and the thought of a little Swiss air is very invigorating.

XXV.

Florence: June 7.

Very hot again, after a good deal of rain in the last two days; or, at least, a good deal must have fallen in the hills, though we only had the tail of the storm here. To-day it is impossible to do anything, even to think; it is a fatigue to exist; and we are eaten up by the mosquitoes.

June 8.

C. has actually achieved a drive, a short one in the Cascine; and we are already feeling half way on our journey to Bologna! The porter and Francesco carried her down, on a little carpet seat, a very clever invention, and we drove off under Carolina's charge, and drank our fresh milk together in the Piazzone. Signor A. invited my father, yesterday, to dine with him at five at the club, where everything was arranged in very good style, a table d'hôte dinner being served to fifteen or twenty gentlemen, each possibly independent of the others.

After the dinner we had one of our pleasant hill drives on the Bologna road, ascending to the left of, and above Fiesole; the sunset beautiful as ever. The corn is beginning to turn golden, and the vines promise well; we have now raspberries and figs; the strawberries are nearly over, and are very dear. We passed near to the city cemetery, where, as at Naples, there are the public pits, one for each day in the year. All the dead bodies are conveyed by night in the dead-carts to this dreadful graveyard from the chapels in or near Florence, to which they are carried by the brothers of the Misericordia.

For many hundred years this Society has been an honoured institution in Italy; founded in the thirteenth century by the *facchini* of Florence, who bound themselves together in works of charity and mercy, filling up the pauses between their daily duties with good deeds; the sick receiving care, and the wounded and dead being carried to their homes. These ancient porters ought, one would fancy, according to orthodox Catholic belief, to have laid up a store of supererogatory virtue, on which the modern herd of *ragazzi* draw rather heavily!

During the great plague of Florence, in the fourteenth century, when the gay and rich inhabitants of the city fled in terror from the pestilence, these Brothers of Mercy, undeterred by the fear of suffering and death, were always at their posts, ready with help and comfort, and by that time many of various ranks and professions had joined the Society, and contributions poured in, proofs of the appreciation or gratitude of their fellow-countrymen; and except for a few years, when the Confraternità was suffered to fall into oblivion, while the funds were appropriated to the service of the pilgrims on their way to and from the Holy Sepulchre, the importance of its services has been recognised by every succeeding Tuscan ruler. Nobles and Grand-dukes, down to the present day, have thought it an honour to be enrolled amongst its members, and its noble charity seems to breathe a pure atmosphere, in which petty distinctions and dissensions, political or religious, are unknown.

The Society is thoroughly well organised; its members are divided into sections, without regard to rank or position, each section consisting of forty individuals, and to all a certain period of active duty is assigned. The tolling of a great bell summons all the brothers who may be on duty at the time to an appointed place of meeting; night or day, there is no appeal, however earnestly occupied at the moment, they never disobey the call; the poor man leaves his work, the citizen his merchandise, the noble his guests or his pleasures; in the dead of night, when the loud peal rouses them from sleep, they must hasten together, and various are the duties they have to perform. There is something wonderfully Christian, *Christ-like*, if I may use the word, in this active personal charity, putting to the blush some of our good Samaritans, who, while they give with princely generosity their 'two pennies' multiplied a hundredfold, would yet shrink from bodily contact with a man who had fallen among thieves.

One of their chief duties is the care of the sick, to whom they act as nurses, and whom they supply with food, medicine, and clothing from their stores, assisting in removing them to the hospital, when required. It is curious to see the ghostly procession suddenly in a street. At first we imagined the burden to be always a dead body, the sufferer is so entirely concealed from view under the black or grey linen or waterproof covering which is extended over an arching wire canopy.

When we were talking a little of attempting to move C. to cooler and quieter rooms, Signor A. said to us: 'You must send for the brothers of the Misericordia;' and on my exclaiming with surprise, added: 'You need not be astonished; that is nothing unusual, I assure you: so skilled are they known to be, and so gentle are their movements, that their services are earnestly requested by the highest ladies and nobles in our city. Their dress entirely concealing their identity, removes all feeling of personal obligation in accepting their help. They are forbidden to receive the smallest gratuity, and may not even break bread in your house, when about the discharge of their duty, even though this should entail prolonged watchings through the night; but, of course, gifts are received by the order.'

They gather together, when summoned by the great bell, in their ancient church in the Piazza del Duomo, where some members of the order are always keeping solemn watch. Here are stored the dresses of the community, litters and biers for the sick and dead, torches, and all things necessary for the performance of the last rites of the church, with a goodly amount of restoratives and necessary medicaments; each litter being provided with a supply of the same, in case of immediate need.

The brothers are also active 'district visitors' (*convisitatori*), the city being divided into sections, each under the care of a certain number of the order, who act as night-watchers, or nurses, as the case may be some women belonging to the Society sharing in this work. There is a *capo guardia*, a sort of captain of the guard, (the different members in turn assuming the office,) and some subordinate officers, whose duty it is to inspect the others, and see that the work committed to each is properly and thoroughly carried out.

The funeral that we saw at Genoa must have been that of a priest in some way connected with their order, as the number of the Confraternità present was much greater than would have been requisite for the ordinary conveyance of a body to the dead-house. The funds of the Society are supplied by the constant contributions given by the people, who are appealed to at all hours by the dark-robed figure whose turn it is to bear the box as *questuante*. These men come suddenly and silently upon you, never speaking, but looking at you through the awful eyeholes of the black hoods, while they shake their *bossolo* entreatingly, acknowledging your gratuity by a quiet bow of thanks. It is pleasant to see with what respect they are everywhere greeted; men standing aside bare-headed when a procession of the long-robed brethren draws near. We, too, looked at them reverently, and bade them God speed in our hearts.

I was nearly stifled last night; my little bed had been hung round with thick white muslin, instead of the ordinary mosquito curtains, and I lay in something very like a cotton box, almost smothered; but that is better than being eaten alive. * * * The least exertion C. ventures on induces an attack of difficulty of breathing. * * *

Last Wednesday, we drove to the Certosa, and beyond it; the grand old monastery, flushing yellow and red in the sunset light, rose above a little cypress wood, and we had a glorious view of distant violet hills, and far-stretching plain. We wound slowly up a steep ascent till we looked down on the convent, and to our right on the misty Val d'Arno, to the left on broad valleys and the highway to Rome. At every little turn of the road there was a perfect picture, with such wonderful colour; at each corner of the wall a tiny shrine, with strong masses of light and shade against the clear blue sky, and vines, the leaves golden-green in the sunlight, and peasants going home from work, and hay-carts drawn by great, soft-eyed, dun-coloured beasts. There was plenty of fresh air, quite a cheery breeze, and we had good horses, who took kindly to the hills.

June 9.

It is a marvel I have survived the night; I had to discard curtains and wooden frame altogether, finding it was not free from very obnoxious insects, though everything about our beds is beautifully clean; this mosquito paraphernalia had been brought from some old store-room. When everything was quieted down again, and the light put out, I heard an ominous buzzing, a faint hum ever drawing nearer; in terror and despair I covered up my head, face and all, with a thick pocket handkerchief, like a pudding prepared for boiling, and then, defying my enemy, lay down to sleep or suffocate. My father is a phrenological study; and on going to a photographer's, I was asked, in reference to a pose, which side of my face was the least bitten!

I have decided to buy a copy of the Cardellino, which Signor Pompignoli is to paint for me from the original, sending the three pictures we have finished together in the same case to England. My share in the matter has been on the school-girl principle of a prize production, finished by the master!*

Dr. W. hopes we may safely start next Wednesday. We owe a great deal to his skill and care, and rather dread, as you may fancy, being thrown on our own resources. * * * * * *

June 11.

It is strange to realise that this will be my last letter from this beautiful city; it seems almost like leaving home again, to think of starting on our travels, and spite of the heat and the mosquitoes, we shall always have tender memories of Florence, and, doubtless, many a haunting desire to return to it some day. We have lived in the shadow of a great sorrow in this lovely southern home, that has made the few weeks we have spent here, a time never to be forgotten; days that lengthened themselves out, crowded with many hopes and fears, and with strength and blessing also, for which we thank God with full hearts.

We are all once more together in our little salon, which is very pleasant, in this sweet Sunday quiet and rest; the dim light coming through green jalousies, and a fresh breeze stirring the muslin curtains; hardly a sound comes up from below, as the busy out-of-door world is taking its siesta, and the church bells are all silent, and the flies and the idlers asleep. Our side-table is loaded with apricots and figs, and a great vase full of magnolias, pinks, Cape-jessamine, and geraniums, whose blossoms scent the air. Carolina's little daughter, Fanny,

---

* Signor Pompignoli, 6, Piazza Santa Croce, is considered the best Raphael copyist in Florence; his charges are very moderate, and I found my commission executed with the greatest care. The picture was sent to MacCracken, his London agents, on the very liberal understanding that if we were disappointed in it, it was to remain in their hands.

is spending the day here, by the landlord's invitation; she looks very sweet, with a modest little face, and fresh, white frock, and has just come up to offer us two of the six bouquets given her after dinner by the couriers, full of happy thanks for some little Gospels, and a Paris brooch, and one of A. M.'s Italian tracts.

No letters to-day, only a newspaper—the 'Times' of Friday, which reaches Florence on Sunday, at two o'clock, *viâ* Mont Cenis.

C. is so much better, growing stronger every day. This morning she had some beef and beans: a real round of beef for a Sunday dinner, just three inches in diameter, made of the *fillet*, rolled.

The weather yesterday was intensely oppressive, and we were glad to rest and read till four. Several strangers were at the dinner table, but the daily numbers are rapidly decreasing; a Milanese family; a young German pair, handsome and golden-haired, on their way to Rome, waiting here for two carriage-bags, containing a large amount of money and all the lady's jewels, which, their maid tells Carolina, they managed to leave in a *voiture* on their way here, and which they are still hoping they may recover; and next to my father a graceful, languid Italian bride and her young husband; she, pretty and charmingly dressed, and always busy with a toothpick, which she amuses herself with biting, while her husband admires her!

We have been with Carolina to the Cascine, the whole long round. Now that the possibility of really accomplishing our journey grows clearer, my father has ceased to care much for our old quiet drives, which were a very good *passe-temps*, in lack of a better. We drank milk, deliciously fresh and frothy, from the Grand Ducal dairy, and watched the carriages and idlers; nurse having the oddest stories always ready about everybody. How she would amuse you! there is such a queer *naïveté* about her histories. We never encourage her, or we should hear the whole stories and adventures of every Florentine we meet, told in English the most piquant, with shrugs and quaint expressions, and illustrations of hands and eyebrows. She is a favourite in the hotel, and comes to our room every fête day, laden with the bouquets given her at the servants' table, and when we say, 'How is it you always have so many, Carolina? Why don't the couriers give any to all the young maids who are there? she answers, with a little comical *moue*, 'Ah, why, indeed; they are young and pretty too, some of them, and they look as if they could eat me, when the couriers come, one after the other, and lay the nosegays on my plate, and say " This is for you, madame," and " Madame, do me the favour to accept my bouquet." Does mademoiselle know we have a proverb in our country, " *Old chickens make good broth?* " '

June 13.

This is only to be a postscript, and a hasty one, too. C. and I have been driving about the town, taking last looks, and buying photographs and straw work, and visiting Fedi's studio. I was determined, if possible, to show C. his great group; so we had both the immense doors flung wide open, and paid a man to turn the marble slowly round, and by the help of a lorgnette she saw it perfectly as she lay in the carriage. My father and I have been to the galleries once more, and our evening's drive all together was very pleasant. We engaged a good basket carriage, large and lounging, with capital horses, which took us up the hill at a great pace.

We drove to Santa Margarita, where I made a little sketch, and the curé's gardener gathered us a last bouquet. The air was pure and cool after the heavy storm on Sunday, the hedges fresh and green, gay with wild roses,

Bologna.

Bologna.

Monsieur Polichinelle !

'traveller's joy,' vines, and Indian corn peeping between the branches. In the fields the crops are all golden and ready for the sickle. We stopped at a little farm to ask for *latte*, and a cow was caught and milked for us by a smiling, good-natured peasant, while his bare-legged sons and daughters watched us with much interest.

XXVI.

Bologna: Thursday, June 15.

* * * We were very busy in many ways that last day in Florence, the *pietra dura* man came again, and we bought some exquisite gold work from him, being tempted at the last moment, and sending Carolina more than once to his little shop. She, receiving our orders with calm philosophy, 'Yes, yes, that is always the way, after my ladies have been for months in the city, they want to buy up half Florence the day they leave, and it is " Carolina, run here," "run there," always the same.'

There were callers to be seen, and cards to be left, and *packing*, and in the evening we had a long charming drive above Fiesole, passing again the Castello Pucci, and the bushes of broom and honeysuckle; as we had set our hearts on sharing its pleasant memories with C. And there we came out into the very teeth of the wind, blowing keen and straight from Vallombrosa, which made us cover up our little patient with cloaks and shawls, and shiver a little at the thought of the Swiss passes.

We did not return to the hotel till half-past eight, to our dismay; and then there were more callers, and the packing to be finished, and somehow the things would not go into the boxes, and C. was sadly tired out before our small world was quiet enough for her to try to go to sleep.

The next morning everything was ready in good time, the kind servants pressing round full of good wishes, the garrulous old chambermaid whom we paid and turned bodily out of the room, and nurse's little daughter waiting to kiss our hands. Carolina came with us to the station, where C. was comfortably established in a carriage and rested quietly with a great bouquet of pinks and pomegranate blossoms, as a last link to the dear City of Flowers. I assure you it was very sad work, except to my father, who was in great spirits at coming away; but even he listened to our talks of another visit here, without saying it was out of the question, which was as much as we could expect under the circumstances, and comforted us a little!

We were off in a hurry at last, with a sad parting from poor Carolina, who clung to us and kissed us, while we cried a little together, and then, with a puff and a snort, we whirled along by the pleasant lanes where we had driven so often on the banks of the Mugnone, and past the Cascine, a green bank of trees on our left, and so up into the hills.

Such a wonderful bit of engineering as this I never saw; we swept round corners, and up steep ascents, and through *forty-five tunnels*, and down rough ravines, that looked like water-courses, across dry beds of rivers, by marvellous zig-zags, and curves, and angles, till we found ourselves in a great plain again, with the old towers of Bologna rising against the sky, and that great chain of the Apennines, up which we had puffed, and through which we had bored and screamed and whistled, and down which we had been whirled and twisted in endless gyrations, lying between us and beloved Florence. C. bore the five hours' journey well, and she did not seem to suffer much, as we had feared she would, from the movement of the carriage or the jolting, which was anything but pleasant as we drove over the little round, rough stones of Bologna to the hotel San Marco, where we found good rooms up a weary flight of stairs, the first-floor being appropriated to kitchens and offices.

Soon after dinner C. went to bed, and I endeavoured to manufacture some arrowroot for her benefit, with an *etna* and spirits of wine. Being unwilling to hurt my feelings she endeavoured to swallow the preparation I presented to her, but I cannot say the cookery was a success; the milk was *heated*, that was something, but the whole concoction was 'lumpy,' and I vainly endeavoured to impress upon her that it was equally nutritious under such circumstances. This morning we went out to explore, leaving C. to rest.

Bologna is the queerest old town; large, and with many handsome buildings. It is a mile one way, by two the other; but all the streets look exactly alike; they are narrow, paved with little round stones, and on either side, broad, shady arcades front every house, sometimes with plain, whitewashed pillars; here and there you pass an old palazzo of dark stone, with quaint carvings, making a pleasant break in the monotony of dull stucco and ugly arches. There are hundreds of these last, with little dark shops underneath, which, last night, looked weird and uncanny, faintly illuminated, with dark figures flitting about the lamps as though busy over some mystery of evil, though they were only worthy citizens, chaffering over the price of a pair of boots; or gossiping about the papers, if the *bottega* were a barber's.

At the end of one street we came upon two rough, square towers of brick, leaning across each other. The Basilica is a grand old building; the square in which it stands was thronged with market people—to me an endless and delightful study. Fancy a real 'Monsieur Polichinelle' exhibiting in the midst of the crowd of stalls to an eager group of gazers, and hitting right and left in puppet wrath, but with a spice of human nature at the bottom, where, veiled and mysterious, stood the man who pulled the strings, with angry thoughts in his heart of the charlatan, 'that good-for-nothing over there, who utters hideous noises, and makes the silly people believe in him —the fools!' There was a vast show of common pottery placed about amongst the unwary feet, which somehow manage to avoid a breakage. A tall man, lean and ill-favoured, and very hungry-looking, had established himself meekly in the rear of the impetuous little marionettes, offering to the on-lookers the harmless refreshment of a water ice, *granito*, for an infinitesimally small coin. I was diverted at watching a very little girl deliberately partaking of the same, with all the air of conscious and becoming dignity that a lady might wear when eating *crème à la vanille*, in the Piazza San Marco at Venice, to the music of the Austrian band.

Last night we walked about under the arcades, and then drove, in a little light calèche, up the hill to San Michele and the Villa Reale, passing crowds of the Bolognese enjoying the delicious evening air. The sunset was very gorgeous, and the rose-coloured light illumined the wide plain extending round us far as the eye could reach, and shone on the old city and its towers. The hedges were green and summer-like, and very English in their luxuriance; and we returned to the hotel much refreshed by our hour's excursion.

To day is the *Fête Dieu*, and a general holiday. Some of the streets were covered with a canvas awning, and hung with crimson draperies, in preparation for the grand procession, which we stopped to see—a most miserable affair: dozens of old men from the workhouse, or from some charitable ecclesiastical establishment; poor, shaking old pensioners, dressed up for the nonce in blue tippets, and carrying tapers in their trembling old hands; five or six choristers chanting, and *one* trumpet, played out of tune; then gilt lanterns on poles, and a great gilt crucifix; and a crowd of Franciscan friars, dirty

Bologna.

Lake Como.

and dismal; and a fresh set of lanterns, and Benedictine monks, looking like gentlemen as compared with their greasy brown brethren. One beautiful pale face I watched for a long time as the procession halted near our carriage; so meek and quiet, with the history of a sealed life under the calm lines of thought, and in the gentle, patient eyes—a man it was sad to see amidst all that low mummery, and those vulgar *confrères*; a great pathetic soul, humbled in the very dust, content to bear the cross of such a life. Through what agonised struggles must it not have passed into that patience, that hourly crucifixion; with a faith true and fervent, though so cruelly mistaken, waiting for the hereafter that should be revealed!

The people of Bologna are far from handsome, and contrast strongly with the Florentines. The women, with little stooping figures and insignificant faces, differ widely from their sisters of Florence, who all walk well and hold their graceful heads high; and those who are really not pretty, make you believe they are good-looking; they have such beautiful eyes and hair, and exquisite taste in their dress, and quiet, dignified movements. Farther north, they are like bundles, badly put together, and shuffle over the ground, while along the Lung' Arno, the women tread like empresses.

We went to one of the churches, the door of which was hidden under the uniform loggia of the street, and visited a chapel of St. Cecilia, and some Francia frescoes, and then drove to the Academia.

The collection of pictures, as all the world knows, is very large, and especially rich in examples of the Bolognese school, many of them being very fine, and marvellous in their colouring; but, grand as the Guidos are—and here they are found in the greatest perfection—I cannot pretend to care for them very much. The great *Pietà* and the Sampson Triumphant, are very wonderful, and I looked at the Massacre of the Innocents with a whole flood of early recollections coming over me of the old engraving in the Pictorial Bible, and own to a pleasure, childish enough in its way, in discovering that that unhappy mother, sitting in the foreground, with clasped hands, wore green velvet sleeves! There is a heavenly Perugino—the Madonna and Child with Four Saints — too beautiful for any words, and one or two Francias worthy to be hung beside it. Raphael's St. Cecilia is gorgeously beautiful, the most brilliant in colour of almost any of his works that I have seen.

I want to read something more about Perugino; I never remember seeing any real history of his life; it ought to have been a good and pure one, like the artist-saint of Fiesole; Perugino paints as one who has lived with the angels, only he must have dwelt in the world too, because the Angelico figures are all etherealised beings, but *his* have suffered and conquered, and look like saints and martyrs, and human souls who have come out of great tribulation, and are rejoicing evermore in a sublime ecstasy of worship.

We visited the prize paintings in the school—the diploma pictures which each student must send in for examination; the successful candidate being supported for a certain number of years, during the completion of his art studies, as in Paris and elsewhere.

It is wonderful to see how utterly and inconceivably bad, modern Italian pictures generally are. We visited an exhibition at Florence, where the walls were hung with daubs, that would often have disgraced a school-girl. Living and breathing as they do in an atmosphere of art, walking daily amidst the grandest paintings of old times, modern Italian painters seem utterly devoid of any powers of real living originality or execution. Those who have any talent among them generally settle down as copyists;

and by hard and incessant work manage to support themselves; but, with one or two exceptions, none move a step further.

Milan, Hôtel de la Ville: Friday.

We came here yesterday afternoon, without stopping at Parma, as C. seemed equal to the longer journey; and it would have been a great pity to have given her the extra fatigue of a night at another hotel, merely to enable us to see the Correggios; so I have put them aside as one of the pleasant things the future may have in store for us. We are comfortably established here in the old quarters, in two rooms on the *entresol*, no others being vacant. While my father dined below, we had a cozy 'high tea' upstairs, with which the waiter brought us a large pint jug full of rich cream! Milan is famous for its dairies, and sends butter both to Florence and to the towns on the Riviera.

C. is getting on well, sleeps well, of course, as she still takes a sedative. She had an attack of difficulty of breathing after the fatigue of dressing; but was able to venture on a short drive with us, when we revisited the grand, beautiful cathedral and Leonardo's *Cenacolo*, which it was interesting to compare with the Raphael and del Sarto frescoes we had just seen, and to feel afresh how preeminent it stands amongst them all, in spite of the havoc time has made.

We drove under the great Napoleon's Arc de Triomphe (M. will remember its painted canvas *double* in Paris, at the entrance there of the army of Italy), and visited his Amphitheatre. We hope to go to Bellaggio to-morrow, and stay there one or two, or three days, as the fancy takes us.

XXVII.

Bellaggio, Hôtel Genazzini, June 18.

HERE we are comfortably established, after a somewhat varied experience. We have capital rooms at the corner looking out on the lake and towards the Swiss mountains, with a broad terrace, which runs round the house, well-supplied with tempting little colonies of tables and chairs. The wind, which has been rising all day, is blowing great guns, tearing and roaring round the house, and dashing the waves into foam against the walls, till one feels quite out at sea, in a gigantic boat, whirled along by the storm. I tried to walk the length of the terrace, and was nearly carried bodily into the water, and came in all in a warm glow from the buffeting. It is delightful after the relaxing heat of Florence. No place could be better for C., who can practise walking on the terrace, and save herself the stairs, and grow strong in the open air. This hotel is full of French and Germans, with one pleasant English family, with whom we have fraternised, but who cross the Splugen to-morrow. There is a long *salle à manger*, like the cabin of a steamer; all glass on one side, with broad, striped, brown awnings that swell out like little sails in the wind, and under them comes a perpetual twinkle of light, reflected on walls and ceiling from the water below. This *salle* opens on a second terrace below and beyond ours—a garden terrace, shady with arcades of Banksia roses and acacias, with endless little tables, at which people breakfast, and write and read, and gossip over their coffee and beer; a flight of steps leads down into the water, where small open boats and large broad tubs, with cushioned seats and gay fringed awnings, rock idly to and fro, waiting for human ballast.

It is rapidly growing dark, the clouds are very stormy, and the lake is a wicked leaden colour, that looks as if it meant mischief; and on the opposite shore a large house has caught fire, its great red angry flames flaring in the wind. We have been watching the black hollows of the windows, the roof is gone and a great tower of smoke is rising up into the pitiless dark night.

My last letter was posted on Saturday, as we were leaving Milan. My father and I visited the Brera, where we were afresh struck by the extreme beauty of some of the frescoes. We drove by the public gardens, and royal palace, where one of the young princes resides, passing through many handsome piazzas, in one of which is the great Cavour statue. He stands smiling down on Milan, the shrewd, powerful face full of earnest thought, a short, stout man, in his ordinary dress, slightly draped in the folds of a loose over-coat, and yet there is something wonderfully grand about the whole, and the bronze figure of Italy at the base writing his name on the marble, is quite beautiful.

Though there may be few modern painters in Italy, there are many sculptors of no mean power. Under the same roof with the collection of miserable pictures in Florence, were rough plaster studies, more finished groups in marble, and designs of rare promise, one I particularly remember, a charmingly fresh model of young Buonarotti, chisel in hand, working at the head of his Faun, the concentrated earnestness in the boy's face most faithfully rendered. During the *festas* at Florence, there were many plaster casts of great citizens, or Greek heroes, set up on wooden pedestals, as decorations in street or piazza; many of great merit, and all ranking above mediocrity, as works of art.

Our railway journey from Milan was easily accomplished, though the train was a slow one, constantly stopping for third class passengers; the scenery, if monotonous, was very pretty. Between Bologna and Milan, near the latter city, you pass fields of rice, and the ground is low and swampy; yesterday the country through which we passed was undulating, and often well wooded, and in the distance there was the range of blue mountains, to tell us we were nearing Switzerland.

At Camerlata we were all turned out into deep white dust to scramble for omnibuses; we secured seats in one filled by an amusing party, and enjoyed our drive to Como, and the oddity of the whole scene, as the train of clumsy vehicles clattered and swung down the long hill, and rattled through the narrow streets of the town.

Our fellow-travellers were a delicate-looking little Italian marquis, in a blue yachting dress and straw hat, and with the largest turquoise I ever saw on one of his fingers; a French gentleman; and another we imagined to be a Syrian, probably a Smyrniote. All three looked as though they might have just returned from a cruise in the eastern waters of the Mediterranean. They had wonderful pink and white boat rugs with them, and the Oriental, a fine bronze specimen, with gleaming eyes and teeth, was gorgeously 'got up'—a tall, lithe man, wearing his flowing draperies with easy grace, dressed in a long white bournous, deep blue tunic and trowsers, crimson-tasseled belt, and a golden-threaded *fichu*, fastened by a gold cord round his head. As we jolted along together in our brightly painted omnibus, with its flapping red blinds, I had an odd fancy that we were a travelling *troupe*, going to act charades somewhere, and dressed for a rehearsal; C. and I having our hats done up in white muslin, our large dust-cloaks and grey dresses, which would admit of any amount of transformation at a moment's notice, the bundle of cloaks and my red blanket being further available 'property.'

We embarked together, but our yachting companions

left us for a beautiful villa on the lake, a little way from Cadenabbia. I must not omit to mention as a striking proof of advanced civilisation, that the London daily penny papers were offered for sale on the steamer, only three days from the date of their publication. There was a pleasant English family on board, with a little Romagnole maid, and a thoroughly English man-servant, of the faithful mastiff order, who looked excessively isolated and uncomfortable in 'foreign parts.' We were tumbled over the side of the steamer into a big flat-bottomed boat, with our old friend *Beppo*, the batellier standing at an oar. Do you remember my sketch of him rowing in a shirt covered over with little demons, and his horror at his general effect in my book, and gasping criticism, 'Ecco il Diavolo!' He has grown very fat, and the black beard is rather grizzled.

We could only find rooms on the ground-floor at the large hotel, and these were so new, and the fresh plaster smelt so unpleasantly, that we had to shift our quarters to some large rooms near the door, which we found, however, so far from attractive, with the incessant noise of passers-by, and no hope of a view of the water, that we moved bodily to this pleasant hotel, where my father has secured the best suite of rooms for us, and now that our hegira is accomplished, we are feeling rested and settled in our charming quarters.

After dinner we walked slowly back to the large hotel,—C. managing the distance very well,—and ascended the staircase to the sloping garden and little chapel. We were almost blown away during the service, and the chaplain looked like a balloon slowly filling, as though he and his surplice must be carried away through the open window beside the pulpit. It quite made me gasp to watch him speaking, with the sense that if he were not careful, he might inhale too much air and finish the sentence in the clouds! The sermon was a very good one ; but the service was a little spoilt to us from anxiety lest C. should suffer, which she does not seem to have done. It was rather like sitting on the top of a lighthouse on a windy evening, with all the glass blown out.

Monday.

This morning, for an hour or two, we were very lazy, watching the beautiful lake and the little boats in the sunshine, strolling up and down the terrace, sitting under the green arcade, till a sudden determination to do *something* seizing us, my father and I started at the very hottest hour of the morning for a little walk, ascending the beautiful gardens of the Villa Serbelloni, and enjoying the glorious views over both arms of the lake. The place is to be sold for 80,000*l.*! I should think they might as well say francs at once, if the family really wishes to part with it. Meanwhile ten gardeners keep the grounds in beautiful order—the house, as you know, is a mere barrack—and are, we imagine, supported by the contributions of visitors. The heat on the higher ground was very great, as we had lost the breeze that seems always to play over the water. * * *

We dined with the five Germans, and I ventured on a little conversation with the Frau step-mother, a pleasant lady, the professor joining in and diverging into good English. Afterwards we went out in a light boat, with one man at the oars. C. thinks she must be rapidly gaining in weight, as he requested my father to sit nearer my side to trim the boat!

I do not know whether we described to you an odd old lady, a Countess G., whom we travelled with to Milan, and afterwards met there, and who followed us to Bellaggio. She had a lively young lady with her, and a very attentive middle-aged gentleman, a sort of *cavaliere servante*, whom I hope she has remembered in her will! Our German friend, who knows everybody, and talks wonderful rhodomontade in any language—speaking

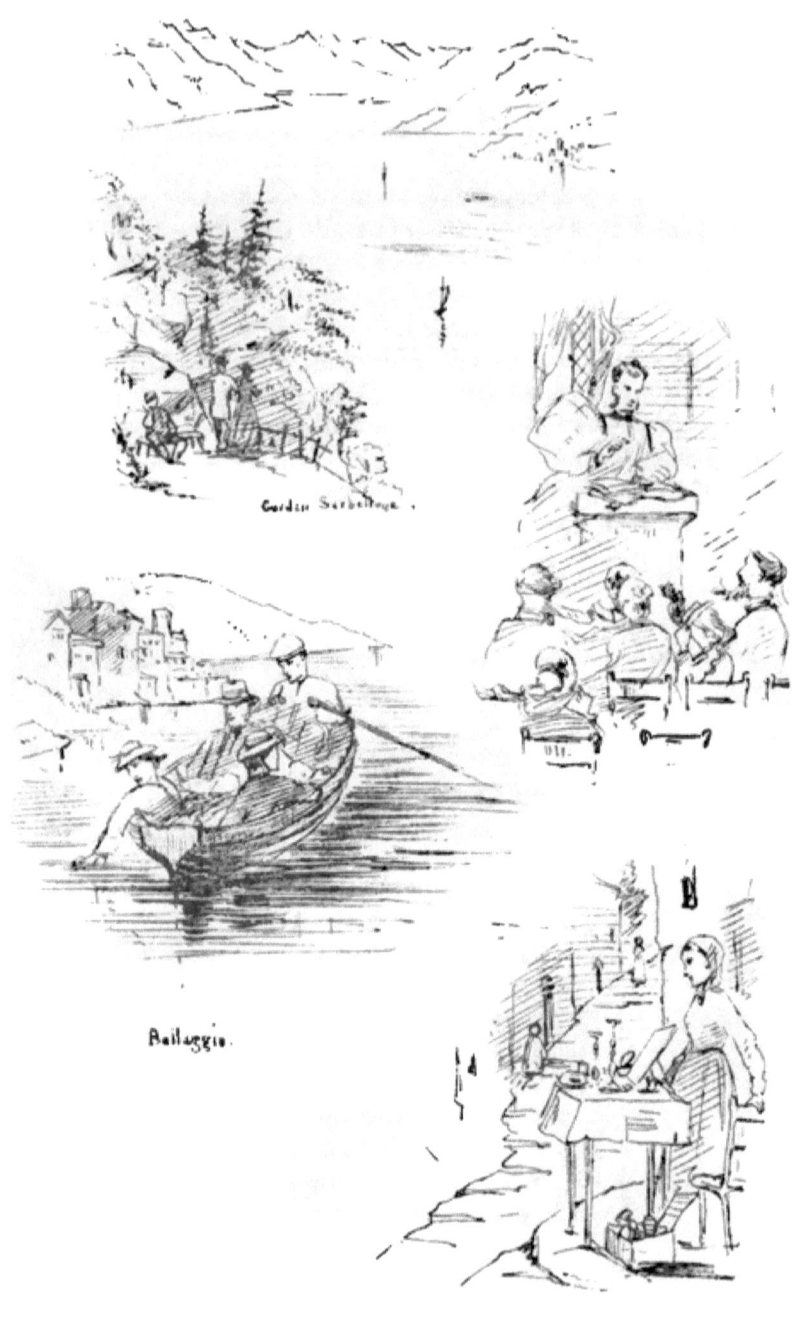

the most amusingly involved English—told my father:
'She is a hundred and fifty, that old lady; she has pretty hands, which she shows; but she wears three veils over her rouge, that she may be mistaken for eighteen!'

Poor old Countess; she was a mournful contrast—in her struggle to grasp the shadow of her youth, and with her strange cracked voice, as its broken echo—to the bright young life beside her; the merry Fraülein, who kept up a gay music of chatter and laughter through the day, and was unwearied in her care of her aged companion, screaming out her little jokes to reach the poor deafened ears, and eking out her explanations with nods and smiles.

I think this must be the identical old lady who always travelled with a passport in which she was described as just three and twenty, and who met the timid suspicions of an unhappy douanier, as to her identity, by the crushing remark:

'Monsieur is the first gentleman who ever questioned my age!'

XXVIII.

Bellaggio: Tuesday, June 20.

WHILST waiting for the table d'hôte, I must try, with some very weak, pale ink, to begin another journal letter. We have not forgotten that this is the anniversary of your wedding day; and very loving thoughts travel homewards, and memory goes pleasantly back to our happy journey together, and the days on this beautiful lake; only now we have the charm of fresh, pure air, and the entire absence of mosquitoes, instead of the intense heat and painful irritation, which certainly did prevent our stay here from being perfect before.

We are enjoying ourselves thoroughly, and the sense of returning health adds to the pleasure; the change has already done me so much good that I am able to walk about and climb the stairs, and to sleep soundly all night. How I wish you were with us again, to share this lovely view of blue hills and dancing, sparkling water, with patches of brilliant colouring everywhere, from boats and awnings and flowers. We shall be very sorry to leave to-morrow; but my father will feel happier when he has taken us safely over the mountains.

This morning we breakfasted at a little rustic table, spread under the shelter of acacias and Banksia rose trees, trained so as to form a thick awning of green leaves, in whose pleasant shade we spend a great part of the day, looking down upon the lake, and listening to the quiet lapping of the water against the steps just below us; watching the little boats passing and re-passing, or throwing crumbs to the tiny fish, which come in crowds to be fed. At a table near us, a Russian lady sometimes sits, her two sweet little children making a play-ground of this cool, pleasant place, and brightening its darkest recesses with their merry faces and laughter. Almost everyone shares in this out-of-door life; and one particular corner has been adopted by a pleasant German gentleman—probably a professor—as his study, where he spends much of his days in writing.

Our boatman of yesterday came to be hired again, and directly after breakfast, the same light, comfortable little boat, with its pretty white awning, was brought round to the steps, and we started for a row across the lake to Cadenabbia. The air was pleasantly fresh, and tempered the heat of the sun, and we went quickly, making almost a little breeze of our own.

My father and E. landed, and walked to the hotel, whilst I stayed behind and floated lazily about, my solitude being enlivened by the arrival of a steamer, and the rather unpleasantly close proximity of her paddles. Amongst the passengers on board were two Englishmen, who, in a great state of indignation and perplexity, appealed to my father to tell them where they were being taken to, as they had no idea of any language but their own, in which they were vainly endeavouring to make themselves understood. The answer to an enquiry as to their desired destination, was: 'We want to go to London,' with a most complete and delightful ignoring of any intermediate stages. As the nearest approach to their demand, they were then being conveyed to Lecco! After explaining to them the most direct homeward route, and paying a pleasant call on our travelling companions of Saturday, who had already sought us out at Bellaggio without finding any clue to our whereabouts, my father and E. rejoined me, and we crossed the lake again to the Villa Melzi, where we landed, and walked about the lovely gardens, full of choice and rare trees and shrubs, and more beautiful than my recollection had painted them.

We enjoyed strolling about, and I am so pleased to find a daily increase in my walking powers. The gardener gave us pomegranate blossoms and a glorious magnolia, and offered us the free use of the grounds, a civility prompted, doubtless, by the promise of an English knife from my father, to be conveyed to him by his brother, the boatman. We saw specimens of most of J.'s pet trees, growing well; and of rare ones, such as the coffee plant, and some flowering shrubs that were quite new to us all. We went into the little chapel, which you will remember; but did not revisit the house.

Our boat brought us back to the hotel, and after buying specimens of the olive wood-work, including some of those useful and ingenious candlesticks that shut into one another for travelling, we came up to our pleasant airy sitting-room, resting on the comfortable sofas with which it is furnished, whilst the mid-day heat made out-of-door life less attractive. Since then we have made a premature descent to the table d'hôte, mistaking the hour, and retreated again to our rooms, after watching my father, with a rod and line, fishing for some runaway papers of the Professor's, which he at last succeeded in landing triumphantly. This gentleman is very pleasant and amusing, a great traveller, and full of wonderful histories of great and small people all over Europe. He was speaking just now of the extreme ignorance of the upper classes in Italy, and says that many of the highest nobles at Rome are unable to sign their names, and he even includes a prince of the Napoleon family in this category.

One anecdote I must tell you, as it gives an idea of the state of society in which such a thing could be done with impunity, which is almost incredible. A Roman Principessa or Marchesa suddenly announced the loss of a very valuable set of diamonds, and demanded the aid of the police in discovering the robbers. Active steps were accordingly taken, and as no one else could reasonably be accused, suspicion fell upon the household, and no one servant appearing more guilty than another, the simplest course, according to the Papal ideas of justice, was to put all the twenty-six in prison, where they remained for eight months. At the end of that time, the missing jewels reappeared as suddenly as they had vanished; the fact becoming well-known that the noble lady, being in some pecuniary difficulties, had sent her diamonds to Russia, in order to raise a large sum of money upon them, and redeemed them when it had suited her to do so. The servants were, doubtless, expected to consider themselves sufficiently compensated for the temporary inconvenience by being restored to their places, or by having their characters cleared from the imputation of dishonesty!

At the table d'hôte, the guests were of a curious description. Near us sat a delicate-looking woman, in a decidedly *négligée* costume, accompanied by an old man, whom we supposed to be her father. She and the Herr Professor had apparently many mutual acquaintances, and he has since told us that she is a Miss ——, a very celebrated photographer at Naples, and considered to be one of the most successful in Europe, but her health has been much injured by the constant use of chemicals, and she is now travelling with her elderly assistant in search of strength and pleasure.

After dinner we had our little boat again, this time without an awning, as the sun had almost dipped behind the hills before we started, and rowed to Varenna, a charmingly picturesque little town on the other side of the Lecco branch of the lake. The evening was very still, and there were not many signs of life amid the quiet houses, some of them bright with masses of greenery and pretty creepers, climbing over balconies or round the pillars of a tiny loggia, up almost amongst the roofs. The road to the Stelvio skirts the lake, and we watched the carts and people passing along it, now in sight, and now disappearing in one of the numberless tunnels cut through the rock. A few bare-footed lads were whipping the lake with their immensely long rods and lines, as we glided close to the shore, whilst in another place some animals had come down to the edge to drink. The sky was cloudless, and the sunset lighted up each range of hills, those nearest us sinking into a deep rich purple, as the twilight shadows began to steal upon us. The evening air was quite cool and fresh, and pleasant as it was to float about, with the quietness and repose over everything, making it an effort to break the silence, prudence prevailed, and our boatmen rowed us briskly to the little landing-place, and everything was arranged for our crossing to Menaggio to-morrow, tea being brought to us, still out-of-doors, at half-past eight o'clock. I have been writing upstairs, but have taken several excursions to the broad terrace outside our windows, to look down upon the group below. My father is making a small illumination with magnesium wire, which is a novelty here, and it is pretty to watch the boats coming through the growing darkness, like moths to a candle, to investigate the mystery.

Porlezza, Lago di Lugano: June 21.

The steamer does not arrive here for nearly an hour, so I can add to my letter whilst we are waiting in this little inn; an elderly Swiss lady and gentleman being our companions in the *salle*. Bellaggio seemed more lovely than ever this morning, and we felt sad at leaving it, and cast longing looks behind us as we were rowed slowly across the lake to Menaggio. We were off about nine o'clock, taking leave before starting of our German friend and of the dear little Russian boy, who delights in watching the boats, and is a great pet with the sailors and everyone in the hotel. The drive from Menaggio here is a very pretty one, but the intense heat lessened our enjoyment of it. There was quite a demand for carriages when we landed, as some other boats reached the village at the same time as ours; and such a wonderful collection of tumble-down vehicles you seldom see, as were ranged for selection in front of the primitive inn, which appears never to have turned its attention to anything beyond supplying means of locomotion to passing travellers. We were amused to watch the packings into queer little chaises, with hat-boxes and portmanteaus slung in all sorts of impossible places, feeling quite distinguished ourselves in a very lofty carriage of rather an antediluvian character, though we were not without suspicions, from their appearance, that the cushions harboured certain ancient predatory tribes, whose energies even a flood would be unable to damp. We contrived to stow away all our luggage in, under, or behind the carriage, and room was also found on the box for a *vetturino*, who is anxious to secure us for the St. Gotthard. We have driven through a continuous garden of chestnuts, walnuts, and mulberries, roses, Indian corn, and wheat; and at Menaggio we bought some of the finest and most deliciously sweet cherries I ever saw. As you descend the hill into this place, you pass the end of the lake, which is very shallow and marshy, suggestive of fever and unhealthiness for the houses near it.

We have found your parcel of magazines most invaluable, and have still the 'Cornhill' and 'Macmillan' reserved for the journey over the mountains.

Schweitzer Hof, Lucerne: Friday, June 23.

I must take up the thread of my journalising from the point where I laid my letter aside, when the steamboat bell summoned us on board. The day had grown a little cloudy, and Lugano altogether looked dull after Como, less bright in colouring, and with less, too, of the southern charm in tinting or scenery; but, from a steamer, one never does appreciate any beauty in the same way as in a smaller boat. Pleasant talk with the old Swiss lady beguiled the way, and in an hour we had reached Lugano, where we landed; finding a comfortable, though old-fashioned inn in the town instead of the Hotel du Parc, which is so far away, and dining and resting, whilst my father inspected the carriage and horses belonging to our *vetturino*, and seeing that they were thoroughly good, arranged with him at once; and at four o'clock we started, a heavy shower coming on at the moment and laying the dust before us, whilst the large thundery drops were quite refreshing after the great heat of the morning. Our man was a native of Mentone, and his regular work was on the Corniche road; but he had been induced to take an English family through North Italy, and whilst waiting for a similar return fare, he was glad to cross and recross the St. Gotthard, though the presence of a stranger in their special locality caused great annoyance and jealousy amongst the *vetturini* on whose preserves he was poaching.

The rain was soon over, but the evening continued cloudy. Our carriage was a very roomy one, with a *coupé*, in which my father established himself, a seat beyond it for the man, and ample space inside for us and all our smaller bags and luggage. The drive over the Monte Cenere was a lovely one, in spite of the absence of sunshine. Sometimes the road wound up-hill, or between meadows full of newly-cut hay, filling the air with its pleasant scent, and amidst which thousands of grasshoppers kept up their ceaseless chirping; and then down into the valley again, by a series of sharp curves and zigzags, the country lying below us like a map, with its white, thread-like roads, and the Ticino flowing through the midst; and all the way the hill-sides were covered with chestnut trees, one mass of great white spikes of bloom, more beautiful in effect than I can describe, and the exquisite fresh green of the young trees by the road side, that had not yet attained to the dignity of much blossom, was very lovely against the background of darker foliage, crowned with its white burden. A glimpse of bright blue water comes unexpectedly at one point of the way, tantalising you with its suggestions of beautiful Maggiore, of which this is the northernmost end.

As we drew nearer Bellinzona, the sky was lighted up with strange orange and copper-coloured clouds, amidst masses of darker ones, and the great hills towering up across the valley, looked blank and stern under their stormy canopy. The air was still, and everything hushed and silent, and we watched for the fulfilment of these thundery portents, but instead of the storm we had anticipated, the ominous clouds began almost insensibly to disperse, and the sky to settle into a calm evening light, giving us every hope of a fine day for the morrow's journey.

We reached our destination by seven o'clock, and being the first arrivals, were fortunate in securing good rooms at the Angelo. Our *vetturino* was anxious we should make a very early start in the morning, but he could not persuade us to be ready before seven o'clock, as we wished to be well rested by a good night in anticipation of the long drive. However, we were punctual to our hour, and the day was a most inviting one, bright, fresh, cool, and inspiriting. The man wanted to be first on the road, and so was rather vexed that our somewhat later start made us third instead. There is a certain pleasurable excitement and amusement in this small emulation, and in constantly meeting the same fellow-travellers, and keeping them in sight all day, halting at the same inns, and comparing notes on what one has seen or done. The carriage which had the much desired lead belonged to a French general, and behind him came an English colonel with his two daughters. A few miles from Bellinzona you stop for additional horses, and there we found the two *vetture* drawn up before the post-house: our *cheval de renfort* was stationed further on in the village, possibly designedly, and to reach it we had to pass the others, and then, to our amusement, the greatest haste was made not to lose one precious moment in the harnessing, and we were off at a great pace, quite infringing all the rules of etiquette on the road, as we found afterwards, our proper course having been to wait solemnly behind the other carriages until they were fairly on their way again. This breach of politeness led to a most violent quarrel between our driver and Colonel C.'s, of which we were to hear more in the course of the day. The French general was travelling post, and soon passed and distanced us again, but our relative positions were otherwise maintained.

s

We enjoyed the varied interests and charms of the drive, and about one o'clock reached Faido,—a pretty little village, where everyone halts for dinner,—being soon followed by the other travellers, including, besides the irate *vetturino* and his charge, a Mr. P. and his young wife and sister, and a delicate-looking lady, travelling homewards from Rome, where her husband is a curate, with 'only five' of her children, mostly boys, a nurse, and governess. The chief thing to be seen at Faido is a fine cascade; but it was beyond my walking powers, so we rested, instead of visiting it, in a clean, comfortable, little room upstairs, after the table d'hôte dinner in the *salle* below. Here we were presently joined by little Mrs. P., who came to ask us about the white muslin cases of our hats, as in a walk with her husband to the waterfall, a sudden gust of wind had left her bareheaded, carrying off her hat, which she was obliged calmly to watch disappear down the foaming torrent—to hasten on its way towards the valley,—whilst she thankfully supplied its place with a very presentable one, found amongst the stores of the village shop, for which she paid tenpence; and having some muslin with her, in a few minutes, with a little cutting and contriving, a really pretty result was obtained.

We lent some of our magazines to the C.'s, who were occupying the next room, and at four o'clock were ready to start again. The other carriage was also waiting, but we chanced to come down first; and then began a regular race, our horses being kept at a tremendous pace, in order that the others, urged on by voice and whip, should not pass us; and all the way a fierce wordy war was carried on between the men (who had both evidently drunk too freely), in spite of any remonstrance or expostulation. We had the best of it, of course, but were very sorry to leave such a legacy of dust to our fellow-travellers, who were enveloped in its clouds almost constantly.

Airolo.

The road was very beautiful, passing through little Alps and pasture lands, or mounting higher and higher above the foaming, dancing Ticino, which kept us company all the way, with constant tributary streams or waterfalls leaping and sparkling down into it from the rocks. The views far and near were lovely, and the sense of being up amongst the mountains very delightful. Our rapid pace still continued, and we dashed on in rather a perilous way sometimes, along a ridge of road ending in a sort of *via mala*, where the whole valley narrows into a channel for the impetuous river; and up above it, under and through the rock, runs the road; and all the time our driver's head was turned from his horses to his enemy, in order to keep the ball of irony and abuse more conveniently going.

We reached Airolo safely, and in about half the usual time from Faido. As we drove among the chalets, and explored the hotel, with its tiny low-roofed rooms and *duvet* furnished beds, we realised that we were at last in Switzerland; and the fresh, exhilarating feeling of the air, which we drank in like iced water, after the heat and dust of the day, was most refreshing and pleasant. As it was still early, E. and I climbed the steep street in search of some pretty subject for a sketch, and were attracted by the sweet smell of newly-cut hay filled with bright Alpine flowers, into a high field, where many peasants were at work, and from whence there was a lovely view in every direction. We returned to our inn before the other travellers arrived, their pace having been much more orthodox; and we watched during the evening with deep sympathy and pity the poor tired lady and her children, who could not be lodged at the *Post*, and could not be fed at the *Tre Re*, so had to put up with the inconvenience of going backwards and forwards between the two homes, the five little things having their tea in our *salle à manger*, and then being carried and led to the other house, and put to bed, all before their mother thought of herself, and came back for her own evening meal.

This morning we were up at about half-past four, and at six o'clock started again in our comfortable carriage, well provided with wraps to keep out the cold, and enjoying the prospect of the long drive. Our *vetturino* we soon found was in rather a dilapidated state and he pleaded guilty to a sprained thumb, whilst his opponent had been seen the evening before with a very bloody face, and we heard afterwards that there had really been a fight between them, in which both were very savagely in earnest. The other man telegraphed to Lugano to tell the authorities to have our driver arrested on his return, but whether he will be punished we shall of course have no means of knowing.

Almost directly after leaving Airolo the steep zig-zags begin; the road wound higher and higher, each fresh curve changing the view, and the whole way by which we had come lying below us, with other carriages creeping round the turnings. The man belonging to our extra horses climbed by pretty paths straight up the banks and through the pine-woods; my father sometimes following his example, and coming back to us with his hands full of lovely flowers in their dewy freshness; amongst them some fine, orange-coloured lilies. Altogether this first part of the ascent reminded us very much of the Splügen. It was all so bright and sparkling in the morning light; the grass everywhere alive with tiny streamlets bubbling and flashing amongst the green, or leaping in miniature cascades from rock to rock. I cannot describe the charm and beauty of it all, and the delightful sense of rising into the purer, clearer, more bracing atmosphere after the debilitating heat of Italy. Suddenly one of the turns in the road brought us in view of a magnificent waterfall in a fine ravine, the chief peculiarity about it being that from that point you see neither the beginning

Crossing the St Gotthard.

The Devil's Bridge.

At the Goldenen Schlüssel.

of the water nor its outlet, and the effect is that of a gigantic spring bursting suddenly from the rocks. It is really one of the sources of the Ticino, numberless small tributaries uniting in this spot, as we saw from a higher part of the road, and flowing downwards through a very narrow fissure or gully quite invisible from the spot where we first noticed the fall.

It was not at all a pleasant change to pass from the sunny brightness and greenness of the higher Alps, far above the pines and chalets, into the gloom and shadow of the loftier, barer mountains, where the snow lay across our road, brown and far from beautiful, and bringing with it a sudden sense of chill and a need for the more vigorous application of all our available wraps. One strongly-built, desolate-looking house stood at the point where this narrow rocky gorge, called the Val Tremola Murray says, from its supposed effect on the nerves of those who pass it, begins. How strange and incomprehensible to us is the life that the inhabitants of this lonely chalet, or refuge, must lead, passed amidst these solitudes, and exposed to every imaginable hardship! The vastness, and dreariness, and silence of everything must have a terrible effect on the mind, one would think, unless from never knowing any brighter, happier life the poor peasant is reduced to the level of his cows, some of whom we watched seeking for pasturage along the almost perpendicular face of the rock opposite to our road, and often slipping and stumbling in their attempts to scramble across the soft snow to a more tempting-looking morsel of grass; a poor bare-footed woman keeping guard over them, and driving or enticing them from the more perilous places. The bitter cold that made us shiver under all our cloaks and rugs, must be endured by this poor creature in all its intensity, with no provision against it in her scanty clothing, her shoeless feet constantly in the snow, and nothing to relieve the horrible monotony of such a life, but the sight of passing travellers, which can have few charms for her.

Traces of the old road are still to be seen, and it must have been even more dreary than the present one just in this spot. The new road is formed by a series of very steep zigzags rising high, one over the other, as far as you can see, and built up in many places with solid masonry: the last turn brings you to more level ground, and before long to the Hospice, a large building, near which is another in course of construction, and a small *albergo*. The sight of men at work, cutting stone and preparing it for the house, gave a pleasant sense of life again, and while we paused for a few minutes to discharge the extra horses and rest the others, we had a little talk with the quiet, kindly mistress of the inn, asking her about her lonely life, and finding a ready acceptance for all the books we had to spare.

Near the Hospice are some small lakes, in one of which the Reuss rises, whilst from the others the Ticino flows southward to join the lower branch we had seen. A succession of very steep zigzags brought us rapidly down to a lower level again, and as we swept round the curves we met sometimes a diligence with its passengers making the ascent on foot, a *vettura*, or a string of heavily laden waggons toiling up the steep road; one of these latter had completely broken down and was lying on one side, fortunately not blocking up the way. The cold was still great and we were not sorry when, about eleven o'clock, we drove into the little village of Hospenthal, passing the old inn which looked deserted and dismal, and halting at the large new one, a thoroughly modern place, full of comforts, and a great improvement on the Goldener Löwe. Here we parted from our *vetturino*, who had evidently some private reason of his own for not wishing to go any farther, and who found for us a very comfortable carriage with good horses to take us the rest of the way; and in the meanwhile we did full justice to the delicious trout, being joined by some of our fellow-travellers at luncheon.

We started again at one o'clock at a good pace, soon reaching Andermatt, and from thence the Devil's Bridge, recalling our walk there with the pleasant Dutch family so many years ago, and our childish impressions of the beauty of the scene, and watching the Reuss tumbling and foaming as noisily and grandly as ever under the slender old archway. We had the pleasant sound and rush of its waters for many a mile, and as we descended more and more into the valley the scene was very lovely, so different from the Italian side of the pass, and yet each beautiful in its own way. Here the pines clustered everywhere, and the bright sunny Alps lay on every hand, with their brown chalets, and the cow-bells tinkling lazily and sweetly in the noonday hush, and then we came to busier life, where the numberless saw-mills took possession of the floating logs to disgorge them again in neat compact piles of boards, and every turn of the road opened out lovely and yet lovelier views of snowy peaks towering up against the deep blue of the sky; and so at last we reached Amstäg, where the horses stopped to rest, and we sat in the cool shady *salle* of the little inn, where you will remember our meeting the lady from Russia, and her little 'Ned,' on their crystal-collecting journey, and how we pitied the poor boy with his heavily laden pockets.

At Altdorf we paid a passing visit to the Goldener Schlüssel, where E. had a rapturous greeting from the landlady and her daughters, who retained a very lively remembrance of their 'zigzag' visit last year. The steamer was waiting at Flüelen, and we had a long, hot voyage, and were very tired before we reached Lucerne; but it was pleasant to find, after being assured by every-one that the Schweitzer Hof was hopelessly full, that my father's letter from Bellaggio had duly arrived, and that Monsieur Hauser had charming rooms ready for us on the first floor, with a little balcony, and a lovely view over the lake. It is a great comfort to be spared an extra flight of steps, as nothing tires me so much as mounting stairs.

I am writing now, on Saturday the 24th, after a good night's rest; and we are none of us the worse for yesterday's fatigues. Since breakfast we have sat in the beautiful reading room, well furnished with easy chairs, sofas, tables, papers, and books; beyond it is the magnificent new *salle*, built since last year, with a conservatory and fountain at the further end, and all decorated in very good taste; it is about sixty feet long, proportionately wide, and thirty feet in height; a small adjoining breakfast-room is also new, the reading-room being the old *salle à manger*. It is difficult to realise that my father remembers the site of this hotel forming part of the lake, with a covered bridge crossing it to the cathedral!

XXIX.

Lucerne: June 25.

WHILST my father and E. have gone to church, I must begin another home letter. The morning is damp, although none of the showers are very heavy; but it seems strange to see rain at all, after the long, unbroken spell of sunshine we have luxuriated in. Yesterday morning we spent almost entirely in our room, enjoying the air and the view from the balcony, with books and work and writing making us pleasantly busy.

Later in the day we went out and walked over the old bridge, returning by the farther one, looking for novelties in the shop windows, and then waiting to see the steamer come in, and recognising amongst her passengers some of our Florence and Bellaggio acquaintances, for whom we had engaged rooms at this hotel. Soon afterwards, as we sat in the balcony, a home voice sounded below, and looking down we saw our English cousins, the C.'s, from whom we had parted last at Florence, after the drive to Santa Margarita. They have been quite adventurous amongst the mountains, and have just come from Interlaken, bringing with them particulars of poor Mrs. Arbuthnot's sad and terrible death on the Schilt Horn, which has excited the most profound interest and sorrow everywhere (as so many had met her on her journey, and been charmed by her grace and sweetness). The account of the sorrowful event is in Friday's 'Times,' which we had here on Saturday.

Many of the faces at the table d'hôte seemed familiar, and amongst them we recognised the American gentleman and his wife, for whom my father prepared a Swiss route at Florence, and it was pleasant to him to find how thoroughly successful it had proved, and how great their enjoyment had been.

We met five or six young men and boys, Swiss or Germans, just come down from the mountains, their hats wreathed with Alpine roses, and carrying poles crowned with large bunches of the bright red blossoms, and green-lined cases and some musical instruments slung across their backs. I think they may possibly appear at the table d'hôte. Yesterday there was a violin and a woman in a red jacket, and a man with a great voice, who sang, and came round with a plate like an *entremet* while we were at dinner.

Tuesday, June 27.

Yesterday morning, after breakfast, we walked to *the Lion*, which is always a delight in its grand simplicity; they have *matched* the old 'Swiss Guard' very well, and dressed in the correct uniform, he no doubt passes muster with many as a veritable survivor of the old band, in spite of the date on the rock.

Here we met the pretty little American lady, who rather startled us as she accosted me with, 'I guess you've had an ill turn,' accompanying the expression of sympathy, with a congratulatory nod on my present amendment.

We stopped on our way home to buy some wood carving, to which I am always faithful, not believing in the theory that similar purchases in England have the same charm for the people who care to possess these small Swiss mementos, and remembering the intense pleasure given to children in the workhouse or in village cottages, and even to their elders, by some tiny figure or bright little chalet that had travelled home in a spare corner of one's luggage, and served to illustrate to the receiver's mind something of what one had seen and done.

E., who had been invited by the C.'s to join them in an expedition to the Righi, made the needful preparations, but some heavy clouds threatening to destroy all hope of a view, the destination was hastily changed for Engelberg, though not in time to save the last boat for Stanz, so that they decided on driving all the way, setting off about twelve o'clock, and we may soon hope for their return.

At two o'clock my father and I went by steamer to Weggis with our American friends, landed, and had just time to see them start on horseback for the Righi, before the return boat carried us back to Lucerne. Directly after the table d'hôte we engaged one of the carriages always waiting in front of the hotel, and had a most pleasant evening drive, crossing the bridge, and after leaving the town behind, passing through a lovely valley with pine woods on the hill sides, and delicious grassy slopes below them, and all the way prosperous looking chalets, farms, and villages. The peasants were carrying the milk to their homes or into Lucerne in the great wooden tubs, all beautifully clean and white. One of the places through which our road lay was almost a town, having a large cotton-mill at work, and boasting a comfortable-looking hotel, a watchmaker's, and a shop with eight silk handkerchiefs tastefully draped in the windows, and some of the houses and gardens were charmingly bright and pretty. Our drive was full of beauty in every turn of the road, and we came back to Lucerne by a way that was quite new to us. We passed a large *caserne* and prison, and met a number of poor men heavily chained, with a guard, returning from work. The town extends a long way on this side, but the houses are for the most part poor, and I fancy strangers rarely come into this quarter. I had tea upstairs, and sat in the balcony till it grew too dark to stay there, finishing the evening with books and work in my room.

Since breakfast this morning we have taken another long drive to Küssnacht, and to a second Tell's chapel beyond it. A carriage from the hotel, with a really magnificent pair of horses, was provided, and we set off about ten o'clock, and drove for some time by the side of the lake, and in front of many villas and *pensions*, with their pretty little boat-houses at the water's edge, turning away presently and mounting a steep hill to a much higher road, lying amongst a continuous orchard of pear and cherry trees—the fruit of both being used to make spirits —passing by sunny little homesteads, with such a wealth of bright flowers in gardens and balconies, and so within sight of the other branch of the lake again, shut in from where we saw it till it seemed to form a complete little basin of its own, the Righi towering up cloudlessly on the opposite side, and Küssnacht, with its cluster of houses and old church tower, at the extreme end.

We drove through the village, and for some little distance beyond it, until we reached the tiny old chapel which commemorates Gessler's death, and is decorated with strange frescoed representations of the scene. Inside it is bare and white-washed, with nothing specially to attract the attention but an old iron box or trunk for alms let into the wall, and so securely locked and padlocked by most complicated contrivances, that you wonder whether the neighbourhood is a dishonest one, or the charity of passers-by great enough to demand such precautions. A few yards further on you come in sight of the Lake of Zug lying far below, and indescribably blue and beautiful. After enjoying the lovely view we turned, retracing our steps to Küssnacht, and ordering some luncheon at the inn, whilst our horses rested. The very good bread, butter and cheese, strawberries and cherries, having been done justice to, we walked out to explore the place, strolling past the village shops and examining their stock-in-trade as displayed in the low windows, ranging, after the manner of the immortal Mr. Tetterby's, from pipes and soap up to a green glass necklace, a Dutch doll, and a pair of stays! We wandered on to the edge of the lake, and watched a group of children on the little landing-stage fishing, noticing that they used poor little grasshoppers for a bait, as in the Engadine.

Our drive home was hardly so enjoyable, as the sun was very powerful, and the enormous flies persisted in buzzing continually round us, as well as annoying the poor horses; but the view in all directions was lovely, and Pilatus, which had been cloudy in the morning, was now perfectly clear; the roadsides were covered with threads of shining gossamer, and the day altogether has been a perfect one for our excursion.

E. has just returned, and must add a postscript to my letter to tell you of her pleasant time amongst the mountains; so good night!

* * * * * *

I have had a day at Engelberg again, a pleasant glimpse at our old Swiss life, a walk to the cascade, with a small barefooted child as guide, who sang *canzonettes* and mountain songs with jodeling chorusses, waking up all sorts of pleasant echoes in one's heart. Old Titlis would not show himself, and was sulkily veiled with clouds both evening and morning; but the other peaks dressed themselves in a golden glow which became them mightily, and all the snow slopes flushed into a smile, and the higher pines raised their heads lovingly into the sunlight, as the soft evening shadows crept round their feet, wrapping woods and hills in a purple covering.

In the Kloster the monks were busy behind their gilded screen, chanting prayers, with whispered responses over their beads from the women, who knelt in silent rows, with folded kerchief, and hair rolled away in the great white coils fastened with large glittering pins; and, as the shadows lengthened, little troops of pedestrians, and botanizing parties, and general idlers, gathered together into the great old pension and round long tables for the *Abendessen*. There were the same little overworked waiter and busy Mädchen, who gave me a most smiling welcome, with earnest enquiries for my last year's companions.

We drove down in the morning under the shelter of the thick beech woods, where the earth is soft with moss, and brown with last year's leaves, through which ground-ivy spreads itself into a dark carpet, studded with bright spots of wood sorrel, with its delicate white blossoms, and sprays of yellow fern, and tangled tracery of flowers; and where the sun, when the wind stirs the leaves, looks down between them, painting them in light and shadow on the pattern below, and bringing out bright red gleams and purple depths from every twisted old trunk, gnarled and moss-grown, whose stems stretch up as pillars, supporting the arches of greenery above. And down far below, hundreds of feet beneath the spot on which you stand, there is more golden light, and just a little gleam of blue, where the summer sky is mirrored in the dancing waters of the torrent that every now and then gets caught among the stones, and makes a quiet side-pool where the sunbeams may find the sky again.

In the broadening plain there were orchards of fruit-trees—apple and walnut—and the picturesque brown cottages with clustering flowers, little village chapels rich with quaint old towers, and a town whose great church gorgeously decorated, looked queerly out of place amidst the strength of the hills and the calm peacefulness of the valley.

The beautiful lake lay resting under the afternoon shadows, lengthening from the mountains, as we drove along its edge, listening to the cool, pleasant lap of the water against rocks and stones; and so the day ended with a gossip over our adventures with our companions, and plannings for our journey on the morrow.

This is my third visit, in as many years, to Engelberg! but I should not have gone now, had not our cousin's kind invitation been too tempting to be refused.

XXX.

Wednesday, June 28. In the Salle d'Attente, Basle.

I SHALL hardly send off another letter before we are in England again. Here we are now, fairly *en route* for home. We dined at the 'Trois Rois,' and had, as usual, a mountainous *omelette soufflée*. It is a great interest to C. and me to see how much larger they become from year to year; and to-day the cook exceeded himself, as you may judge by my sketch. We had the great room to ourselves, except one corner; where a pompous little bristly Englishman and his large wife were dining with 'Arf a bottle of Ockeimer.'

The day is perfect; a cloudless sky and brilliant

sunshine making the country look its brightest, and the river a clear, beautiful green, as it rushed along beneath the hotel balcony.

We were very sorry to leave Lucerne, and our many pleasant old and new friends there; one of the pleasantest, the American sculptor, Mr. Mozier, who has given me a photograph of his last work, Milton's Divine Melancholy; a lovely woman, quite exquisite in face and attitude, and the expression of the upturned head.

Paris: Thursday evening.

We came on through the night; the train was a long one, and not crowded, and the guard promised us a compartment to ourselves, so we fared luxuriously. At one of the stations, tickets were handed to us, on which we read a notice in French, kindly translated for the benefit of the English, as follows:—' Buffet de la Gare à Vesoul. The express and mail trains, which do not stop long at Vesoul; M. M. the travellers who wish to breakfast or to dine, are advertised that they will find at the buffet hat meals in baskets for 2fr. 50c. These meals are composed of three dishes, half a bottle of wine, bread and dessert. M. M. the travellers have thirty minutes to take their meals in their waggons.' Of course such an invitation was irresistible to our curiosity, so we telegraphed for 'three hat dinners in our waggons,' and went on to Vesoul, where we were supplied with the promised crates, or high baskets; and very absurd we must have looked, had there been any one to see, as the train rushed on, while we three sat in a row, each with our unwieldy dumb waiter before us.

They have a very bad arrangement at the Hotel du Louvre, which ought to be changed: an entirely separate day and night service, so that, though our rooms had been ordered from Lucerne, and were actually ready for us, this could not be ascertained till the officials in the office were relieved at seven o'clock; and meanwhile we had to take possession of others for only a couple of hours, to their and our great inconvenience.

C. does not seem the least tired by the night journey.

* * * The honest *cocher* brought our missing waterproofs to this hotel, just after we had started for Lyons. We have had four months of sunshine, and have never missed them!

London: Great Western Hotel.

We watched with some amusement, and a good deal of commiseration, a distrait and unhappy-looking man, sitting forlorn and lonely in the great *salle* of the Louvre Hotel, while we were there. In the morning, when we were starting in one of the little omnibuses for the railway station, *en route* for London, the secretary proposed that another passenger should join us, and a portly figure was stowed away in one corner, who, with many apologies for crowding us, took a quiet survey of his surroundings, whilst we, on our side, had a nearer view of our companion.

A broad-shouldered man, in a suit of shining broadcloth, a dress coat and trousers, all of black, loose about the knees and elbows, a large limp open shirt collar turned back like a sailor's, and a silk handkerchief tied in a knot; a broad, good-humoured face, the large mouth twisting all over into dimples, like a great merry boy's; an eye with an indescribable twinkle, and a wonderful mixture of innocence and shrewdness. A clever 'cute man, with a large heart, simple as a child's.

His hat was planted far back on his head, and in one hand he carried a small portmanteau. His first words were—'I'm a stranger here, I don't understand a word of their language, and I don't like their ways. I manage to get along somehow; I hold out my hand full of money, and say, "help yourselves," and I generally find they're oncommonly well satisfied. But I like to know what I'm about, and so I'm going up to London for a day or two to buy cloaks and things for my little gal to home. I'm from Kentucky, I am,"—and his voice trembled as he told, with a touch of honest pathos, of the sufferings of the border States—'Where you daren't give your own brother a drink of water if he were perishing at your door, if he had been serving down South.' He talked with shrewd common sense of the present state of America, sadly too, like a man who had seen and suffered much.

When we drew up, and were all bundled out at the *embarcadère*, our large shadow still followed us, glad to be assisted through the difficulties of procuring a ticket, and to be steered safely on to the platform. I was carrying my roll of cloaks, which he instantly seized, saying—

'Excuse me, Ma'am, but we don't kinder like to see ladies carrying parcels in America.'

My remonstrances were useless, and I had to allow him to burden himself with our possessions. We were soon settled in a carriage, with our Kentuckian friend in one corner. He handed his credentials to my father, letters of introduction to Mr. Bigelow, the American minister in Paris, and to other well-known names. The one to Mr. Bigelow stated that 'Mr. F. Benjamin Thims, from Lexington, Kentucky, was a real gentleman, of considerable means, who was travelling in *U-rope* for pleasure merely, and for the benefit of his health and spirits, which had been much affected by the recent loss of his wife.'

After reading the papers, my father returned them to their owner, who whispered, with an indescribably comic face, and a wink of his eye—

'I'm to be married again next September; it was all settled just before I come away.'

We said we had noticed him at the Louvre, and thought he seemed rather dull.

'Wall, you see, Ma'am, I was raised to keep Sabbath, and I don't like the ways of these French; they're friv'lous. I am a Presbyterian, and I don't understand them, or their language, or their manners. Now, when I was coming from America, I spent two or three days in London, and I sent my family a history of England in a letter, and now, I suppose they'll be looking out for a history of France, and I'm puzzled how to write it. The ladies come sweeping about with their flounces and their feathers, and the men are hanging reound all day; they're friv'lous, that's what they are, and I don't think anything I can say of them will do them at home much good to read.'

As he asked our advice in his difficulty, I ventured to suggest that, as he had only spent one or two wet days in the great *salle* of the Hôtel du Louvre, and seen the *grandes eaux* play at Versailles, he was perhaps hardly qualified as yet to criticise with much weight the whole French nation; and we had a little talk about the *ouvriers* and their habits, and the country people and the cultivation of their lands, &c., to which he listened very patiently, saying at last—

'I think you're right, Ma'am; there may be more in 'em than I thought, and I'll wait a bit before I write. But for all that, if I was bringing my little gal to Europe, I wouldn't settle in Paris; her mother raised her pious, and to go reg'larly to Sunday school, and I shouldn't like her to see Paris, but I'd bring her to England.'

We talked of the numbers of his countrymen and women who make a home of the French capital, for at least a few months, and are always so wonderfully gay there.

'Yes, I know a good many of 'em: they come over and cut up fine shakes this side of the water. I'm a plain man, and I shouldn't somehow care to do anything, while I was out, that my old mother and my little gal mightn't see.'

On the conversation turning to America again and the

state of the negroes, he told us : 'My mother and I have got a number of them ; we let them go, but I can assure you, they all came back again ; but I guess in places they're sloughing off pretty considerable.'

He had not much affection to spare for the new President. My father, alluding to the disgraceful scene that took place when he was attempting to make his first speech as Vice-President, said he had understood from many of his countrymen that Mr. Johnson had taken stimulants, only by the advice of his doctor, being in poor health at the time.

This elicited another wink from Mr. Thims, who 'calculated he'd been ill, then, *upwards of four years*, to his certain knowledge !'

We met again at Folkestone, and travelled together to London, the Kentuckian looking with keen and appreciating eyes at that fair garden through which we journeyed. On parting, he said : 'I'm a plain man, and a rough one ; but if ever you are in Lexington, just you ask for F. Benjamin Thims, and I'll make you heartily welcome.'

Which we certainly will do, if we ever go there.

From Folkestone to London,
July.

And so this is our very last letter for a long time to come, for which you may be thankful, as writing accomplished in a train is not very legible ; one sight of your faces, and all we shall say together in the first half-hour of meeting will seem more than the many pages our weary pens have travelled over to make you realise our existence day by day, while we have been parted.

We have accomplished the crossing in brilliant sunshine, a blue sea and sky, with just a few waves and little white cloudlets to make us feel the world was awake, and that we were floating back to England. We were very quiet, trying to bring back pictures of all the lovely Southern land we had been journeying through—of the wide plains dotted with houses, like a soft green and purple mantle sprinkled with thousands of seed pearls, and with an embroidered edge of bright coloured flowers and vine tendrils; of that long sea-road, marvellous in its beauty, with the great ocean sleeping in the sunshine, the waves moving gently with the beatings of its heart, and lapping the shore with a little ripple of music over the pebbles, just loud enough to send the water babies to sleep:

> 'Quiet, monotonous, breathless, almost drown'd,
> Inaudibly audible, felt, scarce heard, cometh the sound,
> Monotonous, so monotonous, but oh! so sweet, so sweet,
> When my hid heart is throbbing forth a voice,
>   And the two voices meet.' *

Old Italian cities; galleries rich in art; long ranges of hill and mountain sleeping in the sunset; grey olive trees, and great chestnut trees bright with flowers, and the rows of tall cypresses rising almost black against the golden skies;—we saw them all again as the steamer panted on its way, and as we drew near the shore, the cheery ring of English voices roused us to the realities of the present. A most comfortable ballast are these same realities, sorely needed often by poor humanity, especially when in the form of a woman. Even Mary Stuart could hardly have finished her 'adieu, belle France!' and might possibly have entered Scotland in a better frame of mind, if she had had to look after her own umbrellas, or the packing-case for her guitar!

We were quickly in the train, and off again—homewards. It is worth going away to realise the delight of coming back to England; the dear old country looked so beautiful in its summer dress and sunshine;—

* R. Buchanan.

'Not a grand Nature. Not my chesnut-woods
Of Vallombrosa, cleaving by the spurs
To the precipices. Not my headlong leaps
Of waters, that cry out for joy or fear
In leaping through the palpitating pines,
Like a white soul tossed out to eternity
With thrills of time upon it. Not indeed
My multitudinous mountains, sitting in
The magic circle, with the mutual touch
Electric, panting from their full deep hearts
Beneath the influent heavens, and waiting for
Communion and commission. Italy
Is one thing, England one.
    On English ground
You understand the letter—ere the fall
How Adam lived in a garden. All the fields
Are tied up fast with hedges, nosegay-like;
The hills are crumpled plains, the plains parterres,
The trees round, woolly, ready to be clipped,
And if you seek for any wilderness
You find, at best, a park. A nature tamed
And grown domestic like a barn-door fowl,
Which does not awe you with its claws and beak
Nor tempt you to an eyrie too high up,
But which, in cackling, sets you thinking of
Your eggs to-morrow at breakfast, in the pause
Of finer meditation.
    Rather say,
A sweet familiar nature, stealing in
As a dog might, or child, to touch your hand
Or pluck your gown, and humbly mind you so
Of presence and affection, excellent
For inner uses, from the things without.'

And, after all, what better can we want? Some air is too exciting to be breathed for long, though a little of it is very good to stir us into stronger life, but one comes back with a gasp of content to a more every-day existence.

'Not a grand Nature,' one can dare to say it, thankfully too; how few would really desire to live up to the requirements of such, be it material only, or human! We also, 'in the pauses of finer meditation,' are thinking of 'the fresh eggs,' the 'morrow's breakfast,' and the dear faces round the table. We have had a feast of Nature and Art, both most beautiful, rich and nourishing to the spirit of man; but home love, which satisfies the heart, is worth them all.

How long the moments seem till we meet you once more, and our blue-eyed 'Chrysocoma,' with her sweet child kisses!

By the same Authoress.

HOW WE SPENT THE SUMMER;

OR, A

'VOYAGE EN ZIGZAG,'

WITH SOME MEMBERS OF THE ALPINE CLUB.

'The little bits of landscape scenery—the picturesque cottages and *châlets*—the mountain passes, and lake cities—are the clever jottings of a master hand. ......Looking at the book as the production of an amateur artist, to call it simply clever is not giving it the meed of praise it deserves.......Every page shows a full appreciation of the humorous and pathetic.'—READER.

'It consists of a considerable number of clever and spirited little drawings, rapidly made, by a hand of no mean skill.'—ATHENÆUM.

'A series of very clever and amusing sketches.'—MORNING POST.

'The words "Second Edition" on the cover of this artistic *jeu d'esprit* necessarily denotes a previous appearance of the work, though of this we have no recollection.......Much cannot be said in praise of the sketches, which are everywhere weak as drawings, and often pointless......Still the book is amusing as far as it goes. The idea is not novel; it has been before carried out—and in a far better manner—in a work published a few years ago—*the title of which we do not at the moment remember* (!)'—ART JOURNAL.

'This book is an unmixed delight, except of course that it has a last page. ......There is no class of tourists to which it does not appeal.'

PALL MALL GAZETTE.

London: LONGMANS, GREEN, and CO.

39 Paternoster Row, E.C.
London, *July* 1865.

GENERAL LIST OF WORKS,

NEW BOOKS AND NEW EDITIONS,

PUBLISHED BY

Messrs. LONGMANS, GREEN, READER, and DYER.

Arts, Manufactures, &c. 11	Knowledge for the Young 20
Astronomy, Meteorology, Popular Geography, &c. 7	Miscellaneous and Popular Metaphysical Works 6
Biography and Memoirs 3	Natural History and Popular Science 8
Chemistry, Medicine, Surgery, and the Allied Sciences 9	Periodical Publications 20
Commerce, Navigation, and Mercantile Affairs 18	Poetry and the Drama 17
	Religious Works 12
Criticism, Philology, &c. 4	Rural Sports, &c. 17
Fine Arts and Illustrated Editions 10	Travels, Voyages, &c. 15
General and School Atlases 19	Works of Fiction 16
Historical Works 1	Works of Utility and General Information 18
Index .... 21—24	

Historical Works.

The History of England from the Fall of Wolsey to the Death of Elizabeth. By JAMES ANTHONY FROUDE, M.A. late Fellow of Exeter College, Oxford.

Vols. I. to IV. the Reign of Henry VIII. Third Edition, 54s.

Vols. V. and VI. the Reigns of Edward VI. and Mary. Second Edition, 28s.

Vols. VII. and VIII. the Reign of Elizabeth, Vols. I. and II. Third Edition, 28s.

The History of England from the Accession of James II. By Lord MACAULAY. Three Editions, as follows.

LIBRARY EDITION, 5 vols. 8vo. £4.

CABINET EDITION, 8 vols. post 8vo. 48s.

PEOPLE'S EDITION, 4 vols. crown 8vo. 16s.

Revolutions in English History. By ROBERT VAUGHAN, D.D. 3 vols. 8vo. 45s.

Vol. I. Revolutions of Race, 15s.

Vol. II. Revolutions in Religion, 15s.

Vol. III. Revolutions in Government, 15s.

An Essay on the History of the English Government and Constitution, from the Reign of Henry VII. to the Present Time. By JOHN EARL RUSSELL. Third Edition, revised, with New Introduction. Crown 8vo. 6s.

The History of England during the Reign of George the Third. By WILLIAM MASSEY, M.P. 4 vols. 8vo. 48s.

The Constitutional History of England, since the Accession of George III. 1760—1860. By THOMAS ERSKINE MAY, C.B. Second Edition. 2 vols. 8vo. 33s.

Historical Studies. I. On some of the Precursors of the French Revolution; II. Studies from the History of the Seventeenth Century; III. Leisure Hours of a Tourist. By HERMAN MERIVALE, M.A. 8vo. 12s. 6d.

Lectures on the History of England. By WILLIAM LONGMAN. Vol. I. from the Earliest Times to the Death of King Edward II. with 6 Maps, a coloured Plate, and 53 Woodcuts. 8vo. 15s.

A Chronicle of England, from B.C. 55 to A.D. 1485; written and illustrated by J. E. DOYLE. With 81 Designs engraved on Wood and printed in Colours by E. EVANS. 4to. 42s.

History of Civilization. By HENRY THOMAS BUCKLE. 2 vols. £1 17s.

VOL. I. *England and France*, Fourth Edition, 21s.

VOL. II. *Spain and Scotland*, Second Edition, 16s.

Democracy in America. By ALEXIS DE TOCQUEVILLE. Translated by HENRY REEVE, with an Introductory Notice by the Translator. 2 vols. 8vo. 21s.

The Spanish Conquest in America, and its Relation to the History of Slavery and to the Government of Colonies. By ARTHUR HELPS. 4 vols. 8vo. £3. VOLS. I. & II. 28s. VOLS. III. & IV. 16s. each.

History of the Reformation in Europe in the Time of Calvin. By J. H. MERLE D'AUBIGNÉ, D.D. VOLS. I. and II. 8vo. 28s. and VOL. III. 12s.

Library History of France, in 5 vols. 8vo. By EYRE EVANS CROWE. VOL. I. 14s. VOL. II. 15s. VOL. III. 18s. VOL. IV. nearly ready.

Lectures on the History of France. By the late Sir JAMES STEPHEN, LL.D. 2 vols. 8vo. 24s.

The History of Greece. By C. THIRLWALL, D.D. Lord Bishop of St. David's. 8 vols. 8vo. £3; or in 8 vols. fcp. 28s.

The Tale of the Great Persian War, from the Histories of Herodotus. By GEORGE W. COX, M.A. late Scholar of Trin. Coll. Oxon. Fcp. 7s. 6d.

Ancient History of Egypt, Assyria, and Babylonia. By the Author of 'Amy Herbert.' Fcp. 8vo. 6s.

Critical History of the Language and Literature of Ancient Greece. By WILLIAM MURE, of Caldwell. 5 vols. 8vo. £3 9s.

History of the Literature of Ancient Greece. By Professor K. O. MÜLLER. Translated by the Right Hon. Sir GEORGE CORNEWALL LEWIS, Bart. and by J. W. DONALDSON, D.D. 3 vols. 8vo. 36s.

History of the Romans under the Empire. By CHARLES MERIVALE, B.D. Chaplain to the Speaker.

CABINET EDITION, 8 vols. post 8vo. 48s.
LIBRARY EDITION, 7 vols. 8vo. £5. 11s.

The Fall of the Roman Republic: a Short History of the Last Century of the Commonwealth. By the same Author. 12mo. 7s. 6d.

The Conversion of the Roman Empire the Boyle Lectures for the year 1864, delivered at the Chapel Royal, Whitehall. By the same. 2nd Edition. 8vo. 8s. 6d.

Critical and Historical Essays contributed to the *Edinburgh Review*. By the Right Hon. Lord MACAULAY.

LIBRARY EDITION, 3 vols. 8vo. 36s.
TRAVELLER'S EDITION, in 1 vol. 21s.
In POCKET VOLUMES, 3 vols. fcp. 21s.
PEOPLE'S EDITION, 2 vols. crown 8vo. 8s.

Historical and Philosophical Essays. By NASSAU W. SENIOR. 2 vols. post 8vo. 16s.

History of the Rise and Influence of the Spirit of Rationalism in Europe. By W. E. H. LECKY, M.A. Second Edition. 2 vols. 8vo. 25s.

The Biographical History of Philosophy, from its Origin in Greece to the Present Day. By GEORGE HENRY LEWES. Revised and enlarged Edition. 8vo. 16s.

History of the Inductive Sciences. By WILLIAM WHEWELL, D.D. F.R.S. Master of Trin. Coll. Cantab. Third Edition. 3 vols. crown 8vo. 24s.

Egypt's Place in Universal History; an Historical Investigation. By C. C. J. BUNSEN, D.D. Translated by C. H. COTTRELL, M.A. With many Illustrations. 4 vols. 8vo. £5 8s. VOL. V. is nearly ready, completing the work.

Maunder's Historical Treasury; comprising a General Introductory Outline of Universal History, and a Series of Separate Histories. Fcp. 10s.

Historical and Chronological Encyclopædia, presenting in a brief and convenient form Chronological Notices of all the Great Events of Universal History. By B. B. WOODWARD, F.S.A. Librarian to the Queen. [*In the press.*

History of the Christian Church, from the Ascension of Christ to the Conversion of Constantine. By E. BURTON, D.D. late Regius Prof. of Divinity in the University of Oxford. Eighth Edition. Fcp. 3s. 6d.

Sketch of the History of the Church of England to the Revolution of 1688. By the Right Rev. T. V. SHORT, D.D. Lord Bishop of St. Asaph. Sixth Edition. Crown 8vo. 10s. 6d.

History of the Early Church, from the First Preaching of the Gospel to the Council of Nicæa, A.D. 325. By the Author of 'Amy Herbert.' Fcp. 4s. 6d.

The English Reformation. By F. C. MASSINGBERD, M.A. Chancellor of Lincoln and Rector of South Ormsby. Third Edition, revised and enlarged. Fcp. 6s.

History of Wesleyan Methodism. By GEORGE SMITH, F.A.S. Fourth Edition, with numerous Portraits. 3 vols. crown 8vo. 7s. each.

Villari's History of Savonarola and of his Times, translated from the Italian by LEONARD HORNER, F.R.S. with the co-operation of the Author. 2 vols. post 8vo. with Medallion, 18s.

Lectures on the History of Modern Music, delivered at the Royal Institution. By JOHN HULLAH, Professor of Vocal Music in King's College and in Queen's College, London. FIRST COURSE, with Chronological Tables, post 8vo. 6s. 6d. SECOND COURSE, on the Transition Period, with 26 Specimens, 8vo. 16s.

Biography and Memoirs.

Letters and Life of Francis Bacon, including all his Occasional Works. Collected and edited, with a Commentary, by J. SPEDDING, Trin. Coll. Cantab. VOLS. I. and II. 8vo. 24s.

Passages from the Life of a Philosopher. By CHARLES BABBAGE, Esq. M.A. F.R.S. &c. 8vo. 12s.

Life of Robert Stephenson, F.R.S. By J. C. JEAFFRESON, Barrister-at-Law, and WILLIAM POLE, F.R.S. Memb. Inst. Civ. Eng. With 2 Portraits and 17 Illustrations. 2 vols. 8vo. 32s.

Life of the Duke of Wellington. By the Rev. G. R. GLEIG, M.A. Popular Edition, carefully revised; with copious Additions. Crown 8vo. with Portrait, 5s.

Brialmont and Gleig's Life of the Duke of Wellington. 4 vols. 8vo. with Illustrations, £2 14s.

Life of the Duke of Wellington, partly from the French of M. BRIALMONT, partly from Original Documents. By the Rev. G. R. GLEIG, M.A. 8vo. with Portrait, 15s.

History of my Religious Opinions. By J. H. NEWMAN, D.D. Being the Substance of Apologia pro Vitâ Suâ. Post 8vo. 6s.

Father Mathew: a Biography. By JOHN FRANCIS MAGUIRE, M.P. Popular Edition, with Portrait. Crown 8vo. 3s. 6d.

Rome; its Rulers and its Institutions. By the same Author. New Edition in preparation.

Memoirs, Miscellanies, and Letters of the late Lucy Aikin; including those addressed to Dr. Channing from 1826 to 1842. Edited by P. H. LE BRETON. Post 8vo. 8s. 6d.

Life of Amelia Wilhelmina Sieveking, from the German. Edited, with the Author's sanction, by CATHERINE WINKWORTH. Post 8vo. with Portrait, 12s.

Louis Spohr's Autobiography. Translated from the German. 8vo. 14s.

Felix Mendelssohn's Letters from *Italy and Switzerland,* and *Letters from* 1833 to 1847, translated by Lady WALLACE. New Edition, with Portrait. 2 vols. crown 8vo. 5s. each.

Diaries of a Lady of Quality, from 1797 to 1844. Edited, with Notes, by A. HAYWARD, Q.C. Post 8vo. 10s. 6d.

Recollections of the late William Wilberforce, M.P. for the County of York during nearly 30 Years. By J. S. HARFORD, F.R.S. Second Edition. Post 8vo. 7s.

Memoirs of Sir Henry Havelock, K.C.B. By JOHN CLARK MARSHMAN. Second Edition. 8vo. with Portrait, 12s. 6d.

Thomas Moore's Memoirs, Journal, and Correspondence. Edited and abridged from the First Edition by Earl RUSSELL. Square crown 8vo. with 8 Portraits, 12s. 6d.

Memoir of the Rev. Sydney Smith. By his Daughter, Lady HOLLAND. With a Selection from his Letters, edited by Mrs. AUSTIN. 2 vols. 8vo. 28s.

Vicissitudes of Families. By Sir BERNARD BURKE, Ulster King of Arms. FIRST, SECOND, and THIRD SERIES. 3 vols. crown 8vo. 12s. 6d. each.

Essays in Ecclesiastical Biography. By the Right Hon. Sir J. STEPHEN, LL.D. Fourth Edition. 8vo. 14s.

Biographical Sketches. By NASSAU W. SENIOR. Post 8vo. 10s. 6d.

Biographies of Distinguished Scientific Men. By FRANÇOIS ARAGO. Translated by Admiral W. H. SMYTH, F.R.S. the Rev. B. POWELL, M.A. and R. GRANT, M.A. 8vo. 18s.

Maunder's Biographical Treasury: Memoirs, Sketches, and Brief Notices of above 12,000 Eminent Persons of All Ages and Nations. Fcp. 8vo. 10s.

Criticism, Philosophy, Polity, &c.

Papinian: a Dialogue on State Affairs between a Constitutional Lawyer and a Country Gentleman about to enter Public Life. By GEORGE ATKINSON, B.A. Oxon. Serjeant-at-Law. Post 8vo. 5s.

On Representative Government. By JOHN STUART MILL. Third Edition 8vo. 9s. crown 8vo. 2s.

On Liberty. By the same Author. Third Edition. Post 8vo. 7s. 6d. crown 8vo. 1s. 4d.

Principles of Political Economy. By the same. Sixth Edition. 2 vols. 8vo. 30s. or in 1 vol. crown 8vo. 5s.

A System of Logic, Ratiocinative and Inductive. By the same. Fifth Edition. 2 vols. 8vo. 25s.

Utilitarianism. By the same. 2d Edit. 8vo. 5s.

Dissertations and Discussions. By the same Author. 2 vols. 8vo. 24s.

Examination of Sir W. Hamilton's Philosophy, and of the Principal Philosophical Questions discussed in his Writings. By the same Author. 8vo. 14s.

Lord Bacon's Works, collected and edited by R. L. ELLIS, M.A. J. SPEDDING, M.A. and D. D. HEATH. VOLS. I. to V. *Philosophical Works*, 5 vols. 8vo. £4 6s. VOLS. VI. and VII. *Literary and Professional Works*, 2 vols. £1 16s.

Bacon's Essays, with Annotations. By R. WHATELY, D.D. late Archbishop of Dublin. Sixth Edition. 8vo. 10s. 6d.

Elements of Logic. By R. WHATELY, D.D. late Archbishop of Dublin. Ninth Edition. 8vo. 10s. 6d. crown 8vo. 4s. 6d.

Elements of Rhetoric. By the same Author. Seventh Edition. 8vo. 10s. 6d. crown 8vo. 4s. 6d.

English Synonymes. Edited by Archbishop WHATELY. 5th Edition. Fcp. 3s.

Miscellaneous Remains from the Common-place Book of RICHARD WHATELY, D.D. late Archbishop of Dublin. Edited by Miss E. J. WHATELY. Post 8vo. 7s. 6d.

Essays on the Administrations of Great Britain from 1783 to 1830. By the Right Hon. Sir G. C. LEWIS, Bart. Edited by the Right Hon. Sir E. HEAD, Bart. 8vo. with Portrait, 15s.

By the same Author.

A Dialogue on the Best Form of Government, 4s. 6d.

Essay on the Origin and Formation of the Romance Languages, 7s. 6d.

Historical Survey of the Astronomy of the Ancients, 15s.

Inquiry into the Credibility of the Early Roman History, 2 vols. 30s.

On the Methods of Observation and Reasoning in Politics, 2 vols. 28s.

Irish Disturbances and Irish Church Question, 12s.

Remarks on the Use and Abuse of some Political Terms, 9s.

On Foreign Jurisdiction and Extradition of Criminals, 2s. 6d.

The Fables of Babrius, Greek Text with Latin Notes, PART I. 5s. 6d. PART II. 3s. 6d.

Suggestions for the Application of the Egyptological Method to Modern History, 1s.

An Outline of the Necessary Laws of Thought: a Treatise on Pure and Applied Logic. By the Most Rev. W. THOMSON, D.D. Archbishop of York. Crown 8vo. 5s. 6d.

The Elements of Logic. By Thomas Shedden, M.A. of St. Peter's Coll. Cantab. 12mo. 4s. 6d.

Analysis of Mr. Mill's System of Logic. By W. Stebbing, M.A. Fellow of Worcester College, Oxford. 12mo. 3s. 6d.

The Election of Representatives, Parliamentary and Municipal; a Treatise. By Thomas Hare, Barrister-at-Law. Third Edition, with Additions. Crown 8vo. 6s.

Speeches of the Right Hon. Lord Macaulay, corrected by Himself. 8vo. 12s.

Lord Macaulay's Speeches on Parliamentary Reform in 1831 and 1832. 16mo. 1s.

A Dictionary of the English Language. By R. G. Latham, M.A. M.D. F.R.S. Founded on the Dictionary of Dr. S. Johnson, as edited by the Rev. H. J. Todd, with numerous Emendations and Additions. Publishing in 36 Parts, price 3s. 6d. each, to form 2 vols. 4to.

Thesaurus of English Words and Phrases, classified and arranged so as to facilitate the Expression of Ideas, and assist in Literary Composition. By P. M. Roget, M.D. 14th Edition, crown 8vo. 10s. 6d.

Lectures on the Science of Language, delivered at the Royal Institution. By Max Müller, M.A. Taylorian Professor in the University of Oxford. First Series, Fourth Edition, 12s. Second Series, 18s.

The Debater; a Series of Complete Debates, Outlines of Debates, and Questions for Discussion. By F. Rowton. Fcp. 6s.

A Course of English Reading, adapted to every taste and capacity; or, How and What to Read. By the Rev. J. Pycroft, B.A. Fourth Edition, fcp. 5s.

Manual of English Literature, Historical and Critical: with a Chapter on English Metres. By Thomas Arnold, B.A. Post 8vo. 10s. 6d.

Southey's Doctor, complete in One Volume. Edited by the Rev. J.W. Warter, B.D. Square crown 8vo. 12s. 6d.

Historical and Critical Commentary on the Old Testament; with a New Translation. By M. M. Kalisch, Ph. D. Vol. I. *Genesis*, 8vo. 18s. or adapted for the General Reader, 12s. Vol. II. *Exodus*, 15s. or adapted for the General Reader, 12s.

A Hebrew Grammar, with Exercises. By the same. Part I. *Outlines with Exercises*, 8vo. 12s. 6d. Key, 5s. Part II. *Exceptional Forms and Constructions*, 12s. 6d.

A Latin-English Dictionary. By J. T. White, M.A. of Corpus Christi College, and J. E. Riddle, M.A. of St. Edmund Hall, Oxford. Imp. 8vo. pp. 2,128, 42s.

A New Latin-English Dictionary, abridged from the larger work of *White* and *Riddle* (as above), by J. T. White, M.A. Joint-Author. Medium 8vo. pp. 1,048, 18s.

A Diamond Latin-English Dictionary, or Guide to the Meaning, Quality, and Accentuation of Latin Classical Words. By J. E. Riddle, M.A. 32mo. 2s. 6d.

An English-Greek Lexicon, containing all the Greek Words used by Writers of good authority. By C. D. Yonge, B.A. Fifth Edition. 4to. 21s.

Mr. Yonge's New Lexicon, English and Greek, abridged from his larger work (as above). Square 12mo. 8s. 6d.

A Greek-English Lexicon. Compiled by H. G. Liddell, D.D. Dean of Christ Church, and R. Scott, D.D. Master of Balliol. Fifth Edition, crown 4to. 31s. 6d.

A Lexicon, Greek and English, abridged from Liddell and Scott's *Greek-English Lexicon*. Eleventh Edition, square 12mo. 7s. 6d.

A Practical Dictionary of the French and English Languages. By L. Contanseau. 8th Edition, post 8vo. 10s. 6d.

Contanseau's Pocket Dictionary, French and English, abridged from the above by the Author. New Edition. 18mo. 5s.

New Practical Dictionary of the German Language; German-English, and English-German. By the Rev. W. L. Blackley, M.A., and Dr. Carl Martin Friedlander. Post 8vo. [*In the press.*

Miscellaneous Works and Popular Metaphysics.

Recreations of a Country Parson: being a Selection of the Contributions of A. K. H. B. to *Fraser's Magazine*. SECOND SERIES. Crown 8vo. 3s. 6d.

The Commonplace Philosopher in Town and Country. By the same Author. Crown 8vo. 3s. 6d.

Leisure Hours in Town; Essays Consolatory, Æsthetical, Moral, Social, and Domestic. By the same. Crown 8vo. 3s. 6d.

The Autumn Holidays of a Country Parson: Essays contributed to *Fraser's Magazine* and to *Good Words*, by the same. Crown 8vo. 3s. 6d.

The Graver Thoughts of a Country Parson, SECOND SERIES. By the same. Crown 8vo. 3s. 6d.

Critical Essays of a Country Parson, selected from Essays contributed to *Fraser's Magazine*, by the same. Post 8vo. 9s.

A Campaigner at Home. By SHIRLEY, Author of 'Thalatta,' and 'Nugæ Criticæ.' Post 8vo. with Vignette, 7s. 6d.

Friends in Council: A Series of Readings and Discourses thereon. 2 vols. fcp. 9s.

Friends in Council, SECOND SERIES. 2 vols. post 8vo. 14s.

Essays written in the Intervals of Business. Fcp. 2s. 6d.

Lord Macaulay's Miscellaneous Writings.
LIBRARY EDITION, 2 vols. 8vo. Portrait, 21s.
PEOPLE'S EDITION, 1 vol. crown 8vo. 4s. 6d.

The Rev. Sydney Smith's Miscellaneous Works; including his Contributions to the *Edinburgh Review*.
LIBRARY EDITION, 3 vols. 8vo. 36s.
TRAVELLER'S EDITION, in 1 vol. 21s.
In POCKET VOLUMES, 3 vols. fcp. 21s.
PEOPLE'S EDITION, 2 vols. crown 8vo. 8s.

Elementary Sketches of Moral Philosophy, delivered at the Royal Institution. By the same Author. Fcp. 7s.

The Wit and Wisdom of the Rev. SYDNEY SMITH; a Selection of the most memorable Passages in his Writings and Conversation. 16mo. 7s. 6d.

The History of the Supernatural in All Ages and Nations, and in All Churches, Christian and Pagan; demonstrating a Universal Faith. By WILLIAM HOWITT. 2 vols. post 8vo. 18s.

The Superstitions of Witchcraft. By HOWARD WILLIAMS, M.A. St. John's Coll. Camb. Post 8vo. 7s. 6d.

Chapters on Mental Physiology. By Sir HENRY HOLLAND, Bart. M.D. F.R.S. Second Edition. Post 8vo. 8s. 6d.

Essays selected from Contributions to the *Edinburgh Review*. By HENRY ROGERS. Second Edition. 3 vols. fcp. 21s.

The Eclipse of Faith; or, a Visit to a Religious Sceptic. By the same Author. Tenth Edition. Fcp. 5s.

Defence of the Eclipse of Faith, by its Author; a Rejoinder to Dr. Newman's Reply. Third Edition. Fcp. 3s. 6d.

Selections from the Correspondence of R. E. H. Greyson. By the same Author. Third Edition. Crown 8vo. 7s. 6d.

Fulleriana, or the Wisdom and Wit of THOMAS FULLER, with Essay on his Life and Genius. By the same Author. 16mo. 2s. 6d.

The Secret of Hegel: being the Hegelian System in Origin, Principle, Form, and Matter. By JAMES HUTCHISON STIRLING. 2 vols. 8vo. 28s.

An Introduction to Mental Philosophy, on the Inductive Method. By J. D. MORELL, M.A. LL.D. 8vo. 12s.

Elements of Psychology, containing the Analysis of the Intellectual Powers. By the same Author. Post 8vo. 7s. 6d.

Sight and Touch: an Attempt to Disprove the Received (or Berkeleian) Theory of Vision. By THOMAS K. ABBOTT, M.A. Fellow and Tutor of Trin. Coll. Dublin. 8vo. with 21 Woodcuts, 5s. 6d.

The Senses and the Intellect. By ALEXANDER BAIN, M.A. Prof. of Logic in the Univ. of Aberdeen. Second Edition. 8vo. 15s.

The Emotions and the Will, by the same Author; completing a Systematic Exposition of the Human Mind. 8vo. 15s.

On the Study of Character, including an Estimate of Phrenology. By the same Author. 8vo. 9s.

Time and Space: a Metaphysical Essay. By SHADWORTH H. HODGSON. 8vo. pp. 588, price 16s.

Hours with the Mystics: a Contribution to the History of Religious Opinion. By ROBERT ALFRED VAUGHAN, B.A. Second Edition. 2 vols. crown 8vo. 12s.

Psychological Inquiries. By the late Sir BENJ. C. BRODIE, Bart. 2 vols. or SERIES, fcp. 5s. each.

The Philosophy of Necessity; or, Natural Law as applicable to Mental, Moral, and Social Science. By CHARLES BRAY. Second Edition. 8vo. 9s.

The Education of the Feelings and Affections. By the same Author. Third Edition. 8vo. 3s. 6d.

Christianity and Common Sense. By Sir WILLOUGHBY JONES, Bart. M.A. Trin. Coll. Cantab. 8vo. 6s.

Astronomy, Meteorology, Popular Geography, &c.

Outlines of Astronomy. By Sir J. F. W. HERSCHEL, Bart. M.A. Seventh Edition, revised; with Plates and Woodcuts. 8vo. 18s.

Arago's Popular Astronomy. Translated by Admiral W. H. SMYTH, F.R.S. and R. GRANT, M.A. With 25 Plates and 358 Woodcuts. 2 vols. 8vo. £2 5s.

Arago's Meteorological Essays, with Introduction by Baron HUMBOLDT. Translated under the superintendence of Major-General E. SABINE, R.A. 8vo. 18s.

Saturn and its System. By RICHARD A. PROCTOR, B.A. late Scholar of St. John's Coll. Camb. and King's Coll. London. 8vo. with 14 Plates, 14s.

The Weather-Book; a Manual of Practical Meteorology. By Rear-Admiral ROBERT FITZ ROY, R.N. F.R.S. Third Edition, with 16 Diagrams. 8vo. 15s.

Saxby's Weather System, or Lunar Influence on Weather. By S. M. SAXBY, R.N. Instructor of Naval Engineers. Second Edition. Post 8vo. 4s.

Dove's Law of Storms considered in connexion with the ordinary Movements of the Atmosphere. Translated by R. H. SCOTT, M.A. T.C.D. 8vo. 10s. 6d.

Celestial Objects for Common Telescopes. By T. W. WEBB, M.A. F.R.A.S. With Map of the Moon, and Woodcuts. 16mo. 7s.

Physical Geography for Schools and General Readers. By M. F. MAURY, LL.D. Fcp. with 2 Charts, 2s. 6d.

A Dictionary, Geographical, Statistical, and Historical, of the various Countries, Places, and principal Natural Objects in the World. By J. R. M'CULLOCH. With 6 Maps. 2 vols. 8vo. 63s.

A General Dictionary of Geography, Descriptive, Physical, Statistical, and Historical; forming a complete Gazetteer of the World. By A. KEITH JOHNSTON, F.R.S.E. 8vo. 31s. 6d.

A Manual of Geography, Physical, Industrial, and Political. By W. HUGHES, F.R.G.S. Prof. of Geog. in King's Coll. and in Queen's Coll. Lond. With 6 Maps. Fcp. 7s. 6d.

The Geography of British History; a Geographical Description of the British Islands at Successive Periods. By the same. With 6 Maps. Fcp. 8s. 6d.

Abridged Text-Book of British Geography. By the same. Fcp. 1s. 6d.

The British Empire; a Sketch of the Geography, Growth, Natural and Political Features of the United Kingdom, its Colonies and Dependencies. By CAROLINE BRAY. With 5 Maps. Fcp. 7s. 6d.

Colonisation and Colonies: a Series of Lectures delivered before the University of Oxford. By HERMAN MERIVALE, M.A. Prof. of Polit. Econ. 8vo. 18s.

Maunder's Treasury of Geography, Physical, Historical, Descriptive, and Political. Edited by W. HUGHES, F.R.G.S. With 7 Maps and 16 Plates. Fcp. 10s.

Natural History and Popular Science.

The Elements of Physics or Natural Philosophy. By NEIL ARNOTT, M.D. F.R.S. Physician Extraordinary to the Queen. Sixth Edition. PART I. 8vo. 10s. 6d.

Heat Considered as a Mode of Motion. By Professor JOHN TYNDALL, F.R.S. LL.D. Second Edition. Crown 8vo. with Woodcuts, 12s. 6d.

Volcanos, the Character of their Phenomena, their Share in the Structure and Composition of the Surface of the Globe, &c. By G. POULETT SCROPE, M.P. F.R.S. Second Edition. 8vo. with Illustrations, 15s.

A Treatise on Electricity, in Theory and Practice. By A. DE LA RIVE, Prof. in the Academy of Geneva. Translated by C. V. WALKER, F.R.S. 3 vols. 8vo. with Woodcuts, £3 13s.

The Correlation of Physical Forces. By W. R. GROVE, Q.C. V.P.R.S. Fourth Edition. 8vo. 7s. 6d.

The Geological Magazine; or, Monthly Journal of Geology. Edited by HENRY WOODWARD, F.G.S. F.Z.S. British Museum; assisted by Professor J. MORRIS, F.G.S. and R. ETHERIDGE, F.R.S.E. F.G.S. 8vo. price 1s. monthly.

A Guide to Geology. By J. PHILLIPS, M.A. Prof. of Geol. in the Univ. of Oxford. Fifth Edition; with Plates and Diagrams. Fcp. 4s.

A Glossary of Mineralogy. By H. W. BRISTOW, F.G.S. of the Geological Survey of Great Britain. With 486 Figures. Crown 8vo. 12s.

Phillips's Elementary Introduc-tion to Mineralogy, with extensive Alterations and Additions, by H. J. BROOKE, F.R.S. and W. H. MILLER, F.G.S. Post 8vo. with Woodcuts, 18s.

Van Der Hoeven's Handbook of ZOOLOGY. Translated from the Second Dutch Edition by the Rev. W. CLARK, M.D. F.R.S. 2 vols. 8vo. with 24 Plates of Figures, 60s.

The Comparative Anatomy and Physiology of the Vertebrate Animals. By RICHARD OWEN, F.R.S. D.C.L. 2 vols. 8vo. with upwards of 1,200 Woodcuts.
[In the press.

Homes without Hands: an Account of the Habitations constructed by various Animals, classed according to their Principles of Construction. By Rev. J. G. WOOD, M.A. F.L.S. Illustrations on Wood by G. Pearson, from Drawings by F. W. Keyl and E. A. Smith. In 20 Parts, 1s. each.

Manual of Corals and Sea Jellies. By J. R. GREENE, B.A. Edited by the Rev. J. A. GALBRAITH, M.A. and the Rev. S. HAUGHTON, M.D. Fcp. with 39 Woodcuts, 5s.

Manual of Sponges and Animalculæ; with a General Introduction on the Principles of Zoology. By the same Author and Editors. Fcp. with 16 Woodcuts, 2s.

Manual of the Metalloids. By J. APJOHN, M.D. F.R.S. and the same Editors. Fcp. with 38 Woodcuts, 7s. 6d.

The Sea and its Living Wonders. By Dr. G. HARTWIG. Second (English) Edition. 8vo. with many Illustrations, 18s.

The Tropical World. By the same Author. With 8 Chromoxylographs and 172 Woodcuts. 8vo. 21s.

Sketches of the Natural History of Ceylon. By Sir J. EMERSON TENNENT, K.C.S. LL.D. With 82 Wood Engravings. Post 8vo. 12s. 6d.

Ceylon. By the same Author. 5th Edition; with Maps, &c. and 90 Wood Engravings. 2 vols. 8vo. £2 10s.

A Familiar History of Birds. By E. STANLEY, D.D. F.R.S. late Lord Bishop of Norwich. Seventh Edition, with Woodcuts. Fcp. 3s. 6d.

Marvels and Mysteries of In-stinct; or, Curiosities of Animal Life. By G. GARRATT. Third Edition. Fcp. 7s.

Home Walks and Holiday Ram-bles. By the Rev. C. A. JOHNS, B.A. F.L.S. Fcp. with 10 Illustrations, 6s.

Kirby and Spence's Introduction to Entomology, or Elements of the Natural History of Insects. Seventh Edition. Crown 8vo. 5s.

Maunder's Treasury of Natural History, or Popular Dictionary of Zoology. Revised and corrected by T. S. Cobbold, M.D. Fcp. with 900 Woodcuts, 10s.

The Treasury of Botany, on the Plan of Maunder's Treasury. By J. Lindley, M.D. and T. Moore, F.L.S. assisted by other Practical Botanists. With 16 Plates, and many Woodcuts from designs by W. H. Fitch. Fcp. [*In the press.*

The Rose Amateur's Guide. By Thomas Rivers. 8th Edition. Fcp. 4s.

The British Flora; comprising the Phænogamous or Flowering Plants and the Ferns. By Sir W. J. Hooker, K.H. and G. A. Walker-Arnott, LL.D. 12mo. with 12 Plates, 14s. or coloured, 21s.

Bryologia Britannica; containing the Mosses of Great Britain and Ireland, arranged and described. By W. Wilson. 8vo. with 61 Plates, 42s. or coloured, £4 4s.

The Indoor Gardener. By Miss Maling. Fcp. with Frontispiece, 5s.

Loudon's Encyclopædia of Plants; comprising the Specific Character, Description, Culture, History, &c. of all the Plants found in Great Britain. With upwards of 12,000 Woodcuts. 8vo. £3 13s. 6d.

Loudon's Encyclopædia of Trees and Shrubs; containing the Hardy Trees and Shrubs of Great Britain scientifically and popularly described. With 2,000 Woodcuts. 8vo. 50s.

Maunder's Scientific and Literary Treasury; a Popular Encyclopædia of Science, Literature, and Art. Fcp. 10s.

A Dictionary of Science, Literature, and Art. Fourth Edition. Edited by W. T. Brande, D.C.L. and George W. Cox, M.A., assisted by gentlemen of eminent Scientific and Literary Acquirements. In 12 Parts, each containing 240 pages, price 5s. forming 3 vols. medium 8vo. price 21s. each.

Essays on Scientific and other subjects, contributed to Reviews. By Sir H. Holland, Bart. M.D. Second Edition. 8vo. 14s.

Essays from the Edinburgh and *Quarterly Reviews;* with Addresses and other Pieces. By Sir J. F. W. Herschel, Bart. M.A. 8vo. 18s.

Chemistry, Medicine, Surgery, and the Allied Sciences.

A Dictionary of Chemistry and the Allied Branches of other Sciences. By Henry Watts, F.C.S. assisted by eminent Contributors. 5 vols. medium 8vo. in course of publication in Parts. Vol. I. 3s. 6d. Vol. II. 26s. and Vol. III. 31s. 6d. are now ready.

Handbook of Chemical Analysis, adapted to the Unitary System of Notation: By F. T. Conington, M.A. F.C.S. Post 8vo. 7s. 6d.—Tables of Qualitative Analysis adapted to the same, 2s. 6d.

A Handbook of Volumetrical Analysis. By Robert H. Scott, M.A. T.C.D. Post 8vo. 4s. 6d.

Elements of Chemistry, Theoretical and Practical. By William A. Miller, M.D. LL.D. F.R.S. F.G.S. Professor of Chemistry, King's College, London. 3 vols. 8vo. £2 13s. Part I. Chemical Physics, Third Edition, 12s. Part II. Inorganic Chemistry, 21s. Part III. Organic Chemistry, Second Edition, 20s.

A Manual of Chemistry, Descriptive and Theoretical. By William Odling, M.B. F.R.S. Lecturer on Chemistry at St. Bartholomew's Hospital. Part I. 8vo. 9s.

A Course of Practical Chemistry, for the use of Medical Students. By the same Author. Second Edition, with 70 new Woodcuts. Crown 8vo. 7s. 6d.

The Diagnosis and Treatment of the Diseases of Women; including the Diagnosis of Pregnancy. By Graily Hewitt, M.D. Physician to the British Lying-in Hospital. 8vo. 16s.

Lectures on the Diseases of Infancy and Childhood. By Charles West, M.D. &c. 5th Edition, revised and enlarged. 8vo. 16s.

Exposition of the Signs and Symptoms of Pregnancy: with other Papers on subjects connected with Midwifery. By W. F. Montgomery, M.A. M.D. M.R.I.A. 8vo. with Illustrations, 25s.

A System of Surgery, Theoretical and Practical, in Treatises by Various Authors. Edited by T. HOLMES, M.A. Cantab. Assistant-Surgeon to St. George's Hospital. 4 vols. 8vo. £4 13s.

Vol. I. General Pathology. 21s.

Vol. II. Local Injuries: Gun-shot Wounds, Injuries of the Head, Back, Face, Neck, Chest, Abdomen, Pelvis, of the Upper and Lower Extremities, and Diseases of the Eye. 21s.

Vol. III. Operative Surgery. Diseases of the Organs of Circulation, Locomotion, &c. 21s.

Vol. IV. Diseases of the Organs of Digestion, of the Genito-Urinary System, and of the Breast, Thyroid Gland, and Skin; with APPENDIX and GENERAL INDEX. 30s.

Lectures on the Principles and Practice of Physic. By THOMAS WATSON, M.D. Physician-Extraordinary to the Queen. Fourth Edition. 2 vols. 8vo. 34s.

Lectures on Surgical Pathology. By J. PAGET, F.R.S. Surgeon-Extraordinary to the Queen. Edited by W. TURNER, M.B. 8vo. with 117 Woodcuts, 21s.

A Treatise on the Continued Fevers of Great Britain. By C. MURCHISON, M.D. Senior Physician to the London Fever Hospital. 8vo. with coloured Plates, 18s.

Anatomy, Descriptive and Surgical. By HENRY GRAY, F.R.S. With 410 Wood Engravings from Dissections. Third Edition, by T. HOLMES, M.A. Cantab. Royal 8vo. 28s.

The Cyclopædia of Anatomy and Physiology. Edited by the late R. B. TODD, M.D. F.R.S. Assisted by nearly all the most eminent cultivators of Physiological Science of the present age. 5 vols. 8vo. with 2,853 Woodcuts, £6 6s.

Physiological Anatomy and Physiology of Man. By the late R. B. TODD, M.D. F.R.S. and W. BOWMAN, F.R.S. of King's College. With numerous Illustrations. VOL. II. 8vo. 25s.

A Dictionary of Practical Medicine. By J. COPLAND, M.D. F.R.S. Abridged from the larger work by the Author, assisted by J.C. COPLAND, M.R.C.S. 1 vol. 8vo. [*In the press.*

Dr. Copland's Dictionary of Practical Medicine (the larger work). 3 vols. 8vo. £5 11s.

The Works of Sir B. C. Brodie, Bart. collected and arranged by CHARLES HAWKINS, F.R.C.S.E. 3 vols. 8vo. with Medallion and Facsimile, 48s.

Autobiography of Sir B. C. Brodie, Bart. printed from the Author's materials left in MS. Fcp. 4s. 6d.

Medical Notes and Reflections. By Sir H. HOLLAND, Bart. M.D. Third Edition. 8vo. 18s.

A Manual of Materia Medica and Therapeutics, abridged from Dr. PEREIRA's *Elements* by F. J. FARRE, M.D. Cantab. assisted by R. BENTLEY, M.R.C.S. and by R. WARINGTON, F.C.S. 1 vol. 8vo. [*In October.*

Dr. Pereira's Elements of Materia Medica and Therapeutics, Third Edition, by A. S. TAYLOR, M.D. and G. O. REES, M.D. 3 vols. 8vo. with Woodcuts, £3 15s.

Thomson's Conspectus of the British Pharmacopœia. Twenty-fourth Edition, corrected and made conformable throughout to the New Pharmacopœia of the General Council of Medical Education. By E. LLOYD BIRKETT, M.D. 18mo. 5s. 6d.

Manual of the Domestic Practice of Medicine. By W. B. KESTEVEN, F.R.C.S.E. Second Edition, thoroughly revised, with Additions. Fcp. 5s.

The Fine Arts, and Illustrated Editions.

The New Testament, illustrated with Wood Engravings after the Early Masters, chiefly of the Italian School. Crown 4to. 63s. cloth, gilt top; or £5 5s. elegantly bound in morocco.

Lyra Germanica; Hymns for the Sundays and Chief Festivals of the Christian Year. Translated by CATHERINE WINKWORTH; 125 Illustrations on Wood drawn by J. LEIGHTON, F.S.A. Fcp. 4to. 21s.

Cats' and Farlie's Moral Emblems; with Aphorisms, Adages, and Proverbs of all Nations: comprising 121 Illustrations on Wood by J. LEIGHTON, F.S.A. with an appropriate Text by R. PIGOT. Imperial 8vo. 31s. 6d.

Bunyan's Pilgrim's Progress: with 126 Illustrations on Steel and Wood by C. BENNETT; and a Preface by the Rev. C. KINGSLEY. Fcp. 4to. 21s.

Shakspeare's Sentiments and Similes printed in Black and Gold and illuminated in the Missal style by HENRY NOEL HUMPHREYS. In massive covers, containing the Medallion and Cypher of Shakspeare. Square post 8vo. 21s.

The History of Our Lord, as exemplified in Works of Art; with that of His Types in the Old and New Testament. By Mrs. JAMESON and Lady EASTLAKE. Being the concluding Series of 'Sacred and Legendary Art;' with 13 Etchings and 281 Woodcuts. 2 vols. square crown 8vo. 42s.

In the same Series, by Mrs. JAMESON.

Legends of the Saints and Martyrs. Fourth Edition, with 19 Etchings and 187 Woodcuts. 2 vols. 31s. 6d.

Legends of the Monastic Orders. Third Edition, with 11 Etchings and 88 Woodcuts. 1 vol. 21s.

Legends of the Madonna. Third Edition, with 27 Etchings and 165 Woodcuts. 1 vol. 21s.

Arts, Manufactures, &c.

Encyclopædia of Architecture, Historical, Theoretical, and Practical. By JOSEPH GWILT. With more than 1,000 Woodcuts. 8vo. 42s.

Tuscan Sculptors, their Lives, Works, and Times. With 45 Etchings and 28 Woodcuts from Original Drawings and Photographs. By CHARLES C. PERKINS. 2 vols. imp. 8vo. 63s.

The Engineer's Handbook; explaining the Principles which should guide the young Engineer in the Construction of Machinery. By C. S. LOWNDES. Post 8vo. 5s.

The Elements of Mechanism. By T. M. GOODEVE, M.A. Prof. of Mechanics at the R. M. Acad. Woolwich. Second Edition, with 217 Woodcuts. Post 8vo. 6s. 6d.

Ure's Dictionary of Arts, Manufactures, and Mines. Re-written and enlarged by ROBERT HUNT, F.R.S., assisted by numerous gentlemen eminent in Science and the Arts. With 2,000 Woodcuts. 3 vols. 8vo. £4.

Encyclopædia of Civil Engineering, Historical, Theoretical, and Practical. By E. CRESY, C.E. With above 3,000 Woodcuts. 8vo. 42s.

Treatise on Mills and Millwork. By W. FAIRBAIRN, C.E. F.R.S. With 18 Plates and 322 Woodcuts. 2 vols. 8vo. 32s.

Useful Information for Engineers. By the same Author. FIRST and SECOND SERIES, with many Plates and Woodcuts. 2 vols. crown 8vo. 10s. 6d. each.

The Application of Cast and Wrought Iron to Building Purposes. By the same Author. Third Edition, with 6 Plates and 118 Woodcuts. 8vo. 16s.

The Practical Mechanic's Journal: An Illustrated Record of Mechanical and Engineering Science, and Epitome of Patent Inventions. 4to. price 1s. monthly.

The Practical Draughtsman's Book of Industrial Design. By W. JOHNSON, Assoc. Inst. C.E. With many hundred Illustrations. 4to. 28s. 6d.

The Patentee's Manual: a Treatise on the Law and Practice of Letters Patent for the use of Patentees and Inventors. By J. and J. H. JOHNSON. Post 8vo. 7s. 6d.

The Artisan Club's Treatise on the Steam Engine, in its various Applications to Mines, Mills, Steam Navigation, Railways, and Agriculture. By J. BOURNE, C.E. Sixth Edition; with 37 Plates and 546 Woodcuts. 4to. 42s.

Catechism of the Steam Engine, in its various Applications to Mines, Mills, Steam Navigation, Railways, and Agriculture. By J. BOURNE, C.E. With 199 Woodcuts. Fcp. 9s. The INTRODUCTION of 'Recent Improvements' may be had separately, with 110 Woodcuts, price 3s. 6d.

Handbook of the Steam Engine, by the same Author, forming a KEY to the Catechism of the Steam Engine, with 67 Woodcuts. Fcp. 9s.

The Theory of War Illustrated by numerous Examples from History. By Lieut.-Col. P. L. MACDOUGALL. Third Edition, with 10 Plans. Post 8vo. 10s. 6d.

Collieries and Colliers: A Handbook of the Law and leading Cases relating thereto. By J. C. FOWLER, Barrister-at-Law, Stipendiary Magistrate. Fcp. 6s.

The Art of Perfumery; the History and Theory of Odours, and the Methods of Extracting the Aromas of Plants. By Dr. PIESSE, F.C.S. Third Edition, with 53 Woodcuts. Crown 8vo. 10s. 6d.

Chemical, Natural, and Physical Magic, for Juveniles during the Holidays. By the same Author. Third Edition, enlarged, with 38 Woodcuts. Fcp. 6s.

The Laboratory of Chemical Wonders: A Scientific Mélange for Young People. By the same. Crown 8vo. 5s. 6d.

Talpa; or, the Chronicles of a Clay Farm. By C. W. HOSKYNS, Esq. With 24 Woodcuts from Designs by G. CRUIKSHANK. 16mo. 5s. 6d.

H.R.H. the Prince Consort's Farms; an Agricultural Memoir. By JOHN CHALMERS MORTON. Dedicated by permission to Her Majesty the QUEEN. With 40 Wood Engravings. 4to. 52s. 6d.

Loudon's Encyclopædia of Agriculture: Comprising the Laying-out, Improvement, and Management of Landed Property, and the Cultivation and Economy of the Productions of Agriculture. With 1,100 Woodcuts. 8vo. 31s. 6d.

Loudon's Encyclopædia of Gardening: Comprising the Theory and Practice of Horticulture, Floriculture, Arboriculture, and Landscape Gardening. With 1,000 Woodcuts. 8vo. 31s. 6d.

Loudon's Encyclopædia of Cottage, Farm, and Villa Architecture and Furniture. With more than 2,000 Woodcuts. 8vo. 42s.

History of Windsor Great Park and Windsor Forest. By WILLIAM MENZIES, Resident Deputy Surveyor. With 2 Maps and 20 Photographs. Imp. folio, £8 8s.

The Sanitary Management and Utilisation of Sewage: comprising Details of a System applicable to Cottages, Dwelling-Houses, Public Buildings, and Towns; Suggestions relating to the Arterial Drainage of the Country, and the Water Supply of Rivers. By the same Author. Imp. 8vo. with 9 Illustrations, 12s. 6d.

Bayldon's Art of Valuing Rents and Tillages, and Claims of Tenants upon Quitting Farms, both at Michaelmas and Lady-Day. Eighth Edition, revised by J. C. MORTON. 8vo. 10s. 6d.

Religious and Moral Works.

An Exposition of the 39 Articles, Historical and Doctrinal. By E. HAROLD BROWNE, D.D. Lord Bishop of Ely. Sixth Edition, 8vo. 16s.

The Pentateuch and the Elohistic Psalms, in Reply to Bishop Colenso. By the same. Second Edition. 8vo. 2s.

Examination Questions on Bishop Browne's Exposition of the Articles. By the Rev. J. GORLE, M.A. Fcp. 3s. 6d.

Five Lectures on the Character of St. Paul; being the Hulsean Lectures for 1862. By the Rev. J. S. HOWSON, D.D. Second Edition. 8vo. 9s.

The Life and Epistles of St. Paul. By W. J. CONYBEARE, M.A. late Fellow of Trin. Coll. Cantab. and J. S. HOWSON, D.D. Principal of Liverpool Coll.

LIBRARY EDITION, with all the Original Illustrations, Maps, Landscapes on Steel, Woodcuts, &c. 2 vols. 4to. 48s.

INTERMEDIATE EDITION, with a Selection of Maps, Plates, and Woodcuts. 2 vols. square crown 8vo. 31s. 6d.

PEOPLE'S EDITION, revised and condensed, with 46 Illustrations and Maps. 2 vols. crown 8vo. 12s.

The Voyage and Shipwreck of St. Paul; with Dissertations on the Ships and Navigation of the Ancients. By JAMES SMITH, F.R.S. Crown 8vo. Charts, 8s. 6d.

A Critical and Grammatical Commentary on St. Paul's Epistles. By C. J. ELLICOTT, D.D. Lord Bishop of Gloucester and Bristol. 8vo.

Galatians, Third Edition, 8s. 6d.

Ephesians, Third Edition, 8s. 6d.

Pastoral Epistles, Third Edition, 10s. 6d.

Philippians, Colossians, and Philemon, Third Edition, 10s. 6d.

Thessalonians, Second Edition, 7s. 6d.

Historical Lectures on the Life of Our Lord Jesus Christ: being the Hulsean Lectures for 1859. By the same Author. Fourth Edition. 8vo. 10s. 6d.

The Destiny of the Creature; and other Sermons preached before the University of Cambridge. By the same. Post 8vo. 5s.

The Broad and the Narrow Way; Two Sermons preached before the University of Cambridge. By the same. Crown 8vo. 2s.

Rev. T. H. Horne's Introduction to the Critical Study and Knowledge of the Holy Scriptures. Eleventh Edition, corrected, and extended under careful Editorial revision. With 4 Maps and 22 Woodcuts and Facsimiles. 4 vols. 8vo. £3 13s. 6d.

Rev. T. H. Horne's Compendious Introduction to the Study of the Bible, being an Analysis of the larger work by the same Author. Re-edited by the Rev. JOHN AYRE, M.A. With Maps, &c. Post 8vo. 9s.

The Treasury of Bible Knowledge, on the plan of Maunder's Treasuries. By the Rev. JOHN AYRE, M.A. Fcp. 8vo. with Maps and Illustrations. [*In the press.*

The Greek Testament; with Notes, Grammatical and Exegetical. By the Rev. W. WEBSTER, M.A. and the Rev. W. F. WILKINSON, M.A. 2 vols. 8vo. £2 4s.

 VOL. I. the Gospels and Acts, 20s.

 VOL. II. the Epistles and Apocalypse, 24s.

The Four Experiments in Church and State; and the Conflicts of Churches. By Lord ROBERT MONTAGU, M.P. 8vo. 12s.

Every-day Scripture Difficulties explained and illustrated; Gospels of St. Matthew and St. Mark. By J. E. PRESCOTT, M.A. 8vo. 9s.

The Pentateuch and Book of Joshua Critically Examined. By the Right Rev. J. W. COLENSO, D.D. Lord Bishop of Natal. People's Edition, in 1 vol. crown 8vo. 6s. or in 5 Parts, 1s. each.

The Pentateuch and Book of Joshua Critically Examined. By Prof. A. KUENEN, of Leyden. Translated from the Dutch, and edited with Notes, by the Right Rev. J. W. COLENSO, D.D. Bishop of Natal. 8vo. 8s. 6d.

The Formation of Christendom. PART I. By T. W. ALLIES. 8vo. 12s.

Christendom's Divisions; a Philosophical Sketch of the Divisions of the Christian Family in East and West. By EDMUND S. FFOULKES, formerly Fellow and Tutor of Jesus Coll. Oxford. Post 8vo. 7s. 6d.

The Life of Christ, an Eclectic Gospel, from the Old and New Testaments, arranged on a New Principle, with Analytical Tables, &c. By CHARLES DE LA PRYME, M.A. Trin. Coll. Camb. Revised Edition. 8vo. 5s.

The Hidden Wisdom of Christ and the Key of Knowledge; or, History of the Apocrypha. By ERNEST DE BUNSEN. 2 vols. 8vo. 28s.

Hippolytus and his Age; or, the Beginnings and Prospects of Christianity. By Baron BUNSEN, D.D. 2 vols. 8vo. 30s.

Outlines of the Philosophy of Universal History, applied to Language and Religion: Containing an Account of the Alphabetical Conferences. By the same Author. 2 vols. 8vo. 33s.

Analecta Ante-Nicæna. By the same Author. 3 vols. 8vo. 42s.

Essays on Religion and Literature. By various Writers. Edited by H. E. MANNING, D.D. 8vo. 10s. 6d.

Essays and Reviews. By the Rev. W. TEMPLE, D.D. the Rev. R. WILLIAMS, B.D. the Rev. B. POWELL, M.A. the Rev. H. B. WILSON, B.D. C. W. GOODWIN, M.A. the Rev. M. PATTISON, B.D. and the Rev. B. JOWETT, M.A. 12th Edition. Fcp. 8vo. 5s.

Mosheim's Ecclesiastical History. MURDOCK and SOAMES's Translation and Notes, re-edited by the Rev. W. STUBBS, M.A. 3 vols. 8vo. 45s.

Bishop Jeremy Taylor's Entire Works: With Life by BISHOP HEBER. Revised and corrected by the Rev. C. P. EDEN, 10 vols. £5 5s.

Passing Thoughts on Religion. By the Author of 'Amy Herbert.' 8th Edition. Fcp. 5s.

Thoughts for the Holy Week, for Young Persons. By the same Author. 3d Edition. Fcp. 8vo. 2s.

Night Lessons from Scripture. By the same Author. 2d Edition. 32mo. 3s.

Self-examination before Confirmation. By the same Author. 32mo. 1s. 6d.

Readings for a Month Preparatory to Confirmation from Writers of the Early and English Church. By the same. Fcp. 4s.

Readings for Every Day in Lent, compiled from the Writings of Bishop JEREMY TAYLOR. By the same. Fcp. 5s.

Preparation for the Holy Communion; the Devotions chiefly from the works of JEREMY TAYLOR. By the same. 32mo. 3s.

Morning Clouds. Second Edition. Fcp. 5s.

Spring and Autumn. By the same Author. Post 8vo. 6s.

The Wife's Manual; or, Prayers, Thoughts, and Songs on Several Occasions of a Matron's Life. By the Rev. W. CALVERT, M.A. Crown 8vo. 10s. 6d.

Spiritual Songs for the Sundays and Holidays throughout the Year. By J. S. B. MONSELL, LL.D. Vicar of Egham. Fourth Edition. Fcp. 4s. 6d.

The Beatitudes: Abasement before God; Sorrow for Sin; Meekness of Spirit; Desire for Holiness; Gentleness; Purity of Heart; the Peace-makers; Sufferings for Christ. By the same. 2d Edition, fcp. 3s. 6d.

Hymnologia Christiana; or, Psalms and Hymns selected and arranged in the order of the Christian Seasons. By B. H. KENNEDY, D.D. Prebendary of Lichfield. Crown 8vo. 7s. 6d.

Lyra Domestica; Christian Songs for Domestic Edification. Translated from the *Psaltery and Harp* of C. J. P. SPITTA, and from other sources, by RICHARD MASSIE. FIRST and SECOND SERIES, fcp. 4s. 6d. each.

Lyra Sacra; Hymns, Ancient and Modern, Odes, and Fragments of Sacred Poetry. Edited by the Rev B. W. SAVILE, M.A. Fcp. 5s.

Lyra Germanica, translated from the German by Miss C. WINKWORTH. FIRST SERIES, Hymns for the Sundays and Chief Festivals; SECOND SERIES, the Christian Life. Fcp. 5s. each SERIES.

Hymns from Lyra Germanica, 18mo. 1s.

Historical Notes to the 'Lyra Germanica;' containing brief Memoirs of the Authors of the Hymns, and Notices of Remarkable Occasions on which some of them have been used; with Notices of other German Hymn Writers. By THEODORE KÜBLER. Fcp. 7s. 6d.

Lyra Eucharistica; Hymns and Verses on the Holy Communion, Ancient and Modern; with other Poems. Edited by the Rev. ORBY SHIPLEY, M.A. Second Edition. Fcp. 7s. 6d.

Lyra Messianica; Hymns and Verses on the Life of Christ, Ancient and Modern; with other Poems. By the same Editor. Fcp. 7s. 6d.

Lyra Mystica; Hymns and Verses on Sacred Subjects, Ancient and Modern. By the same Editor. Fcp. 7s. 6d.

The Chorale Book for England; a complete Hymn-Book in accordance with the Services and Festivals of the Church of England; the Hymns translated by Miss C. WINKWORTH; the Tunes arranged by Prof. W. S. BENNETT and OTTO GOLDSCHMIDT. Fcp. 4to. 12s. 6d.

Congregational Edition. Fcp. 2s.

The Catholic Doctrine of the Atonement; an Historical Inquiry into its Development in the Church: with an Introduction on the Principle of Theological Developments. By H. N. OXENHAM, M.A. formerly Scholar of Balliol College, Oxford. 8vo. 8s. 6d.

From Sunday to Sunday; an attempt to consider familiarly the Weekday Life and Labours of a Country Clergyman. By R. GEE, M.A. Vicar of Abbott's Langley and Rural Dean. Fcp. 5s.

First Sundays at Church; or, Familiar Conversations on the Morning and Evening Services of the Church of England. By J. E RIDDLE, M.A. Fcp. 2s. 6d.

The Judgment of Conscience, and other Sermons. By RICHARD WHATELY, D.D. late Archbishop of Dublin. Crown 8vo. 4s. 6d.

Paley's Moral Philosophy, with Annotations. By RICHARD WHATELY, D.D. late Archbishop of Dublin. 8vo. 7s.

Travels, Voyages, &c.

Outline Sketches of the High Alps of Dauphiné. By T. G. BONNEY, M.A. F.G.S. M.A.C. Fellow of St. John's Coll. Camb. With 13 Plates and a Coloured Map. Post 4to. 16s.

Ice Caves of France and Switzerland; a narrative of Subterranean Exploration. By the Rev. G. F. BROWNE, M A. Fellow and Assistant-Tutor of St. Catherine's Coll. Cambridge, M.A.C. With 11 Woodcuts. Square crown 8vo. 12s. 6d.

Village Life in Switzerland. By SOPHIA D. DELMARD. Post 8vo. 9s. 6d.

How we Spent the Summer; or, a Voyage en Zigzag in Switzerland and Tyrol with some Members of the ALPINE CLUB. From the Sketch-Book of one of the Party. In oblong 4to. with about 300 Illustrations, 10s. 6d.

Map of the Chain of Mont Blanc, from an actual Survey in 1863—1864. By A. ADAMS-REILLY, F.R.G.S. M.A.C. Published under the Authority of the Alpine Club. In Chromolithography on extra stout drawing-paper 28in. × 17in. price 10s. or mounted on canvas in a folding case, 12s. 6d.

The Hunting Grounds of the Old World; FIRST SERIES, Asia. By H. A. L. the Old Shekarry. Third Edition, with 7 Illustrations. 8vo. 18s.

Camp and Cantonment; a Journal of Life in India in 1857—1859, with some Account of the Way thither. By Mrs. LEOPOLD PAGET. To which is added a Short Narrative of the Pursuit of the Rebels in Central India by Major PAGET, R.H.A. Post 8vo. 10s. 6d.

Explorations in South - west Africa, from Walvisch Bay to Lake Ngami and the Victoria Falls. By THOMAS BAINES, F.R.G.S. 8vo. with Maps and Illustrations, 21s.

South American Sketches; or, a Visit to Rio Janeiro, the Organ Mountains, La Plata, and the Paraná. By THOMAS W. HINCHLIFF, M.A. F.R.G.S. Post 8vo. with Illustrations, 12s. 6d.

Vancouver Island and British Columbia; their History, Resources, and Prospects. By MATTHEW MACFIE, F.R.G.S. With Maps and Illustrations. 8vo. 18s.

History of Discovery in our Australasian Colonies, Australia, Tasmania, and New Zealand, from the Earliest Date to the Present Day. By WILLIAM HOWITT. With 3 Maps of the Recent Explorations from Official Sources. 2 vols. 8vo. 28s.

The Capital of the Tycoon; a Narrative of a 3 Years' Residence in Japan. By Sir RUTHERFORD ALCOCK, K.C.B. 2 vols. 8vo. with numerous Illustrations, 42s.

Last Winter in Rome. By C. R. WELD. With Portrait and Engravings on Wood. Post 8vo. 14s.

Autumn Rambles in North Africa. By JOHN ORMSBY, of the Middle Temple. With 16 Illustrations. Post 8vo. 8s. 6d.

The Dolomite Mountains. Excursions through Tyrol, Carinthia, Carniola, and Friuli in 1861, 1862, and 1863. By J. GILBERT and G. C. CHURCHILL, F.R.G.S. With numerous Illustrations. Square crown 8vo. 21s.

A Summer Tour in the Grisons and Italian Valleys of the Bernina. By Mrs. HENRY FRESHFIELD. With 2 Coloured Maps and 4 Views. Post 8vo. 10s. 6d.

Alpine Byways; or, Light Leaves gathered in 1859 and 1860. By the same Authoress. Post 8vo. with Illustrations, 10s. 6d.

A Lady's Tour Round Monte Rosa; including Visits to the Italian Valleys. With Map and Illustrations. Post 8vo. 14s.

Guide to the Pyrenees, for the use of Mountaineers. By CHARLES PACKE. With Maps, &c. and Appendix. Fcp. 6s.

The Alpine Guide. By JOHN BALL, M.R.I.A. late President of the Alpine Club. Post 8vo. with Maps and other Illustrations.

Guide to the Western Alps, including Mont Blanc, Monte Rosa, Zermatt, &c. 7s. 6d.

Guide to the Oberland and all Switzerland, excepting the Neighbourhood of Monte Rosa and the Great St. Bernard; with Lombardy and the adjoining portion of Tyrol. 7s. 6d.

Christopher Columbus; his Life, Voyages, and Discoveries. Revised Edition, with 4 Woodcuts. 18mo. 2s. 6d.

Captain James Cook; his Life, Voyages, and Discoveries. Revised Edition, with numerous Woodcuts. 18mo. 2s. 6d.

Narratives of Shipwrecks of the Royal Navy between 1793 and 1857, compiled from Official Documents in the Admiralty by W. O. S. GILLY; with a Preface by W. S. GILLY, D.D. 3rd Edition, fcp. 5s.

A Week at the Land's End. By J. T. BLIGHT; assisted by E. H. RODD, R. Q. COUCH, and J. RALFS. With Map and 96 Woodcuts. Fcp. 6s. 6d.

Visits to Remarkable Places: Old Halls, Battle-Fields, and Scenes illustrative of Striking Passages in English History and Poetry. By WILLIAM HOWITT. 2 vols. square crown 8vo. with Wood Engravings, 25s.

The Rural Life of England. By the same Author. With Woodcuts by Bewick and Williams. Medium 8vo. 12s. 6d.

Works of Fiction.

Late Laurels: a Tale. By the Author of 'Wheat and Tares.' 2 vols. post 8vo. 15s.

A First Friendship. [Reprinted from *Fraser's Magazine*.] Crown 8vo. 7s. 6d.

Atherstone Priory. By L. N. COMYN. 2 vols. post 8vo. 21s.

Ellice: a Tale. By the same. Post 8vo. 9s. 6d.

Stories and Tales by the Author of 'Amy Herbert,' uniform Edition, each Tale or Story complete in a single volume.

AMY HERBERT, 2s. 6d.	KATHARINE ASHTON, 3s. 6d.
GERTRUDE, 2s. 6d.	MARGARET PERCIVAL, 5s.
EARL'S DAUGHTER, 2s. 6d.	LANETON PARSONAGE, 4s. 6d.
EXPERIENCE OF LIFE, 2s. 6d.	URSULA, 4s. 6d.
CLEVE HALL, 3s. 6d.	
IVORS, 3s. 6d.	

A Glimpse of the World. By the Author of 'Amy Herbert.' Fcp. 7s. 6d.

Essays on Fiction, reprinted chiefly from Reviews, with Additions. By NASSAU W. SENIOR. Post 8vo. 10s. 6d.

Elihu Jan's Story; or, the Private Life of an Eastern Queen. By WILLIAM KNIGHTON, LL.D. Assistant-Commissioner in Oudh. Post 8vo. 7s. 6d.

The Six Sisters of the Valleys: an Historical Romance. By W. BRAMLEY-MOORE, M.A. Incumbent of Gerrard's Cross, Bucks. Third Edition, with 14 Illustrations. Crown 8vo. 5s.

The Gladiators: a Tale of Rome and Judæa. By G. J. WHYTE MELVILLE. Crown 8vo. 5s.

Digby Grand, an Autobiography. By the same Author. 1 vol. 5s.

Kate Coventry, an Autobiography. By the same. 1 vol. 5s.

General Bounce, or the Lady and the Locusts. By the same. 1 vol. 5s.

Holmby House, a Tale of Old Northamptonshire. 1 vol. 5s.

Good for Nothing, or All Down Hill. By the same. 1 vol. 6s.

The Queen's Maries, a Romance of Holyrood. 1 vol. 6s.

The Interpreter, a Tale of the War. By the same. 1 vol. 5s.

Tales from Greek Mythology. By GEORGE W. COX, M.A. late Scholar of Trin. Coll. Oxon. Second Edition. Square 16mo. 3s. 6d.

Tales of the Gods and Heroes. By the same Author. Second Edition. Fcp. 5s.

Tales of Thebes and Argos. By the same Author. Fcp. 4s. 6d.

The Warden: a Novel. By ANTHONY TROLLOPE. Crown 8vo. 3s. 6d.

Barchester Towers: a Sequel to 'The Warden.' By the same Author. Crown 8vo. 5s.

Poetry and the Drama.

Select Works of the British Poets; with Biographical and Critical Prefaces by Dr. AIKIN; with Supplement, of more recent Selections, by LUCY AIKIN. Medium 8vo. 18s.

Goethe's Second Faust. Translated by JOHN ANSTER, LL.D. M.R.I.A. Regius Professor of Civil Law in the University of Dublin. Post 8vo. 15s.

Tasso's Jerusalem Delivered, translated into English Verse by Sir J. KINGSTON JAMES, Kt. M.A. 2 vols. fcp. with Facsimile, 14s.

Poetical Works of John Edmund Reade; with final Revision and Additions. 3 vols. fcp. 18s. or each vol. separately, 6s.

Moore's Poetical Works, Cheapest Editions complete in 1 vol. Including the Autobiographical Prefaces and Author's last Notes, which are still copyright. Crown 8vo. ruby type, with Portrait, 7s. 6d. or People's Edition, in larger type, 12s. 6d.

Moore's Poetical Works, as above, Library Edition, medium 8vo. with Portrait and Vignette, 14s. or in 10 vols. fcp. 3s. 6d. each.

Tenniel's Edition of Moore's *Lalla Rookh,* with 68 Wood Engravings from Original Drawings and other Illustrations. Fcp. 4to. 21s.

Moore's Lalla Rookh. 32mo. Plate, 1s. 16mo. Vignette, 2s. 6d.

Maclise's Edition of Moore's Irish *Melodies,* with 161 Steel Plates from Original Drawings. Super-royal 8vo. 31s. 6d.

Moore's Irish Melodies, 32mo. Portrait, 1s. 16mo. Vignette, 2s. 6d.

Southey's Poetical Works, with the Author's last Corrections and copyright Additions. Library Edition, in 1 vol. medium 8vo. with Portrait and Vignette, 14s. or in 10 vols. fcp. 3s. 6d. each.

Lays of Ancient Rome; with *Ivry* and the *Armada.* By the Right Hon. LORD MACAULAY. 16mo. 4s. 6d.

Lord Macaulay's Lays of Ancient Rome. With 90 Illustrations on Wood, Original and from the Antique, from Drawings by G. SCHARF. Fcp. 4to. 21s.

Poems. By JEAN INGELOW. Ninth Edition. Fcp. 8vo. 5s.

Poetical Works of Letitia Elizabeth Landon (L.E.L.) 2 vols. 16mo. 10s.

Playtime with the Poets: a Selection of the best English Poetry for the use of Children. By a LADY. Crown 8vo. 5s.

Bowdler's Family Shakspeare, cheaper Genuine Edition, complete in 1 vol. large type, with 36 Woodcut Illustrations, price 14s. or, with the same ILLUSTRATIONS, in 6 pocket vols. 3s. 6d. each.

Arundines Cami, sive Musarum Cantabrigiensium Lusus Canori. Collegit atque edidit H. DRURY, M.A. Editio Sexta, curavit H. J. HODGSON, M.A. Crown 8vo. 7s. 6d.

Rural Sports, &c.

Encyclopædia of Rural Sports; a Complete Account, Historical, Practical, and Descriptive, of Hunting, Shooting, Fishing, Racing, &c. By D. P. BLAINE. With above 600 Woodcuts (20 from Designs by JOHN LEECH). 8vo. 42s.

Notes on Rifle Shooting. By Captain HEATON, Adjutant of the Third Manchester Rifle Volunteer Corps. Fcp. 2s. 6d.

Col. Hawker's Instructions to Young Sportsmen in all that relates to Guns and Shooting. Revised by the Author's SON. Square crown 8vo. with Illustrations, 18s.

The Dead Shot, or Sportsman's Complete Guide; a Treatise on the Use of the Gun, Dog-breaking, Pigeon-shooting, &c. By MARKSMAN. Fcp. 8vo. with Plates, 5s.

The Fly-Fisher's Entomology.
By ALFRED RONALDS. With coloured
Representations of the Natural and Artificial Insect. 6th Edition; with 20 coloured
Plates. 8vo. 14s.

Hand-book of Angling : Teaching
Fly-fishing, Trolling, Bottom-fishing, Salmon-fishing; with the Natural History of
River Fish, and the best modes of Catching
them. By EPHEMERA. Fcp. Woodcuts, 5s.

The Cricket Field ; or, the History
and the Science of the Game of Cricket. By
JAMES PYCROFT, B.A. Trin. Coll. Oxon.
4th Edition. Fcp. 5s.

The Cricket Tutor ; a Treatise exclusively
Practical. By the same. 18mo. 1s.

Cricketana. By the same Author. With 7
Portraits of Cricketers. Fcp. 5s.

The Horse: with a Treatise on Draught.
By WILLIAM YOUATT. New Edition, revised and enlarged. 8vo. with numerous
Woodcuts, 10s. 6d.

The Dog. By the same Author. 8vo. with
numerous Woodcuts, 6s.

**The Horse's Foot, and how to keep
it Sound.** By W. MILES, Esq. 9th Edition,
with Illustrations. Imp. 8vo. 12s. 6d.

A Plain Treatise on Horse-shoeing. By
the same Author. Post 8vo. with Illustrations, 2s. 6d.

Stables and Stable Fittings. By the same.
Imp. 8vo. with 13 Plates, 15s.

Remarks on Horses' Teeth, addressed to
Purchasers. By the same. Post 8vo. 1s. 6d.

On Drill and Manœuvres of
Cavalry, combined with Horse Artillery.
By Major-Gen. MICHAEL W. SMITH, C.B.
Commanding the Poonah Division of the
Bombay Army. 8vo. 12s. 6d.

The Dog in Health and Disease.
By STONEHENGE. With 70 Wood Engravings. Square crown 8vo. 15s.

The Greyhound in 1864. By the same
Author. With 24 Portraits of Greyhounds.
Square crown 8vo. 21s.

The Ox, his Diseases and their Treatment; with an Essay on Parturition in the
Cow. By J. R. DOBSON, M.R.C.V.S. Crown
8vo. with Illustrations, 7s. 6d.

Commerce, Navigation, and Mercantile Affairs.

The Law of Nations Considered
as Independent Political Communities. By
TRAVERS TWISS, D.C.L. Regius Professor
of Civil Law in the University of Oxford.
2 vols. 8vo. 30s. or separately, PART I. *Peace*,
12s. PART II. *War*, 18s.

A Nautical Dictionary, defining
the Technical Language relative to the
Building and Equipment of Sailing Vessels
and Steamers, &c. By ARTHUR YOUNG.
Second Edition; with Plates and 150 Woodcuts. 8vo. 18s.

**A Dictionary, Practical, Theoretical, and Historical, of Commerce and
Commercial Navigation.** By J. R. M'CULLOCH. 8vo. with Maps and Plans, 50s.

The Study of Steam and the
Marine Engine, for Young Sea Officers. By
S. M. SAXBY, R.N. Post 8vo. with 87
Diagrams, 5s. 6d.

A Manual for Naval Cadets. By
J. M'NEIL BOYD, late Captain R.N. Third
Edition; with 240 Woodcuts, and 11 coloured
Plates. Post 8vo. 12s. 6d.

Works of Utility and General Information.

Modern Cookery for Private
Families, reduced to a System of Easy
Practice in a Series of carefully-tested
Receipts. By ELIZA ACTON. Newly revised and enlarged; with 8 Plates, Figures,
and 150 Woodcuts. Fcp. 7s. 6d.

The Handbook of Dining ; or, Corpulency and Leanness scientifically considered. By BRILLAT-SAVARIN, Author of
'Physiologie du Goût.' Translated by
L. F. SIMPSON. Revised Edition, with
Additions. Fcp. 3s. 6d.

On Food and its Digestion; an Introduction to Dietetics. By W. BRINTON, M.D. Physician to St. Thomas's Hospital, &c. With 48 Woodcuts. Post 8vo. 12s.

Wine, the Vine, and the Cellar. By THOMAS G. SHAW. Second Edition, revised and enlarged, with Frontispiece and 31 Illustrations on Wood. 8vo. 16s.

A Practical Treatise on Brewing; with Formulæ for Public Brewers, and Instructions for Private Families. By W. BLACK. 8vo. 10s. 6d.

Short Whist. By MAJOR A. The Sixteenth Edition, revised, with an Essay on the Theory of the Modern Scientific Game by PROF. P. Fcp. 3s. 6d.

Whist, What to Lead. By CAM. Second Edition. 32mo. 1s.

Hints on Etiquette and the Usages of Society; with a Glance at Bad Habits. Revised, with Additions, by a LADY of RANK. Fcp. 2s. 6d.

The Cabinet Lawyer; a Popular Digest of the Laws of England, Civil and Criminal. 20th Edition, extended by the Author; including the Acts of the Sessions 1863 and 1864. Fcp. 10s. 6d.

The Philosophy of Health; or, an Exposition of the Physiological and Sanitary Conditions conducive to Human Longevity and Happiness. By SOUTHWOOD SMITH, M.D. Eleventh Edition, revised and enlarged; with 113 Woodcuts. 8vo. 15s.

Hints to Mothers on the Management of their Health during the Period of Pregnancy and in the Lying-in Room. By T. BULL, M.D. Fcp. 5s.

The Maternal Management of Children in Health and Disease. By the same Author. Fcp. 5s.

Notes on Hospitals. By FLORENCE NIGHTINGALE. Third Edition, enlarged; with 13 Plans. Post 4to. 18s.

C. M. Willich's Popular Tables for Ascertaining the Value of Lifehold, Leasehold, and Church Property, Renewal Fines, &c.; the Public Funds; Annual Average Price and Interest on Consols from 1731 to 1861; Chemical, Geographical, Astronomical, Trigonometrical Tables, &c. Post 8vo. 10s.

Thomson's Tables of Interest, at Three, Four, Four and a Half, and Five per Cent., from One Pound to Ten Thousand and from 1 to 365 Days. 12mo. 3s. 6d.

Maunder's Treasury of Knowledge and Library of Reference: comprising an English Dictionary and Grammar, Universal Gazetteer, Classical Dictionary, Chronology, Law Dictionary, Synopsis of the Peerage, useful Tables, &c. Fcp. 10s.

General and School Atlases.

An Atlas of History and Geography, representing the Political State of the World at successive Epochs from the commencement of the Christian Era to the Present Time, in a Series of 16 coloured Maps. By J. S. BREWER, M.A. Third Edition, revised, &c. by E. C. BREWER, LL.D. Royal 8vo. 15s.

Bishop Butler's Atlas of Modern Geography, in a Series of 33 full-coloured Maps, accompanied by a complete Alphabetical Index. New Edition, corrected and enlarged. Royal 8vo. 10s. 6d.

Bishop Butler's Atlas of Ancient Geography, in a Series of 24 full-coloured Maps, accompanied by a complete Accentuated Index. New Edition, corrected and enlarged. Royal 8vo. 12s.

School Atlas of Physical, Political, and Commercial Geography, in 17 full-coloured Maps, accompanied by descriptive Letterpress. By E. HUGHES F.R.A.S. Royal 8vo. 10s. 6d.

Middle-Class Atlas of General Geography, in a Series of 29 full-coloured Maps, containing the most recent Territorial Changes and Discoveries. By WALTER M'LEOD, F.R.G.S. 4to. 5s.

Physical Atlas of Great Britain and Ireland; comprising 30 full-coloured Maps, with illustrative Letterpress, forming a concise Synopsis of British Physical Geography. By WALTER M'LEOD, F.R.G.S. Fcp. 4to. 7s. 6d.

Periodical Publications.

The Edinburgh Review, or Critical Journal, published Quarterly in January, April, July, and October. 8vo. price 6s. each No.

The Geological Magazine, or Monthly Journal of Geology, edited by HENRY WOODWARD, F.G.S.; assisted by Prof. J. MORRIS, F.G.S. and R. ETHERIDGE, F.R.S.E. F.G.S. 8vo. price 1s. each No.

Fraser's Magazine for Town and Country, published on the 1st of each Month. 8vo. price 2s. 6d. each No.

The Alpine Journal: a Record of Mountain Adventure and Scientific Observation. By Members of the Alpine Club. Edited by H. B. GEORGE, M.A. Published Quarterly, May 31, Aug. 31, Nov. 30, Feb. 28. 8vo. price 1s. 6d. each No.

Knowledge for the Young.

The Stepping Stone to Knowledge: Containing upwards of Seven Hundred Questions and Answers on Miscellaneous Subjects, adapted to the capacity of Infant Minds. By a MOTHER. New Edition, enlarged and improved. 18mo. price 1s.

The Stepping Stone to Geography: Containing several Hundred Questions and Answers on Geographical Subjects. 18mo. 1s.

The Stepping Stone to English History: Containing several Hundred Questions and Answers on the History of England. 1s.

The Stepping Stone to Bible Knowledge: Containing several Hundred Questions and Answers on the Old and New Testaments. 18mo. 1s.

The Stepping Stone to Biography: Containing several Hundred Questions and Answers on the Lives of Eminent Men and Women. 18mo. 1s.

Second Series of the Stepping Stone to Knowledge: containing upwards of Eight Hundred Questions and Answers on Miscellaneous Subjects not contained in the FIRST SERIES. 18mo. 1s.

The Stepping Stone to French Pronunciation and Conversation: Containing several Hundred Questions and Answers. By Mr. P. SADLER. 18mo. 1s.

The Stepping Stone to English Grammar; containing several Hundred Questions and Answers on English Grammar. By Mr. P. SADLER. 18mo. 1s.

The Stepping Stone to Natural History: VERTEBRATE OR BACKBONED ANIMALS. PART I. *Mammalia*; PART II. *Birds, Reptiles, Fishes.* 18mo. 1s. each Part.

The Instructor; or, Progressive Lessons in General Knowledge. Originally published under the Direction of the Committee of General Literature and Education of the Society for Promoting Christian Knowledge. 7 vols. 18mo. freely illustrated with Woodcuts and Maps, price 14s.

I. Exercises, Tales, and Conversations on Familiar Subjects; with Easy Lessons from History. Revised and improved Edition. Price 2s.

II. Lessons on Dwelling-Houses and the Materials used in Building Them; on Articles of Furniture; and on Food and Clothing. Revised and improved Edition. Price 2s.

III. Lessons on the Universe; on the Three Kingdoms of Nature, Animal, Vegetable, and Mineral; on the Structure, Senses, and Habits of Man; and on the Preservation of Health. Revised and improved Edition. 2s.

IV. Lessons on the Calendar and Almanack; on the Twelve Months of the Year; and on the appearances of Nature in the Four Seasons, Spring, Summer, Autumn, and Winter. Revised and improved Edition. Price 2s.

V. Descriptive Geography with Popular Statistics of the various Countries and Divisions of the Globe, their People and Productions. Revised and improved Edition. With 6 Maps. 2s.

VI. Elements of Ancient History, from the Formation of the First Great Monarchies to the Fall of the Roman Empire. Revised and improved Edition. Price 2s.

VII. Elements of [Mediæval and] Modern History, from A.D. 406 to A.D. 1862: with brief Notices of European Colonies. Revised and improved Edition. Price 2s.

INDEX.

ABBOTT on Sight and Touch	6
ACTON'S Modern Cookery	18
AIKIN'S Select British Poets	17
———— Memoirs and Remains	3
ALCOCK'S Residence in Japan	15
ALLIES on Formation of Christianity	13
Alpine Guide (The)	15
———— Journal (The)	20
APJOHN'S Manual of the Metalloids	8
ARAGO'S Biographies of Scientific Men	4
———— Popular Astronomy	7
———— Meteorological Essays	7
ARNOLD'S Manual of English Literature	5
ARNOTT'S Elements of Physics	8
Arundines Cami	17
Atherstone Priory	16
ATKINSON'S Papinian	4
Autumn Holidays of a Country Parson	6
AYRE'S Treasury of Bible Knowledge	13
BABBAGE'S Life of a Philosopher	3
BACON'S Essays, by WHATELY	4
———— Life and Letters, by SPEDDING	3
———— Works, by ELLIS, SPEDDING, and HEATH	4
BAIN on the Emotions and Will	6
———— on the Senses and Intellect	6
———— on the Study of Character	6
BAINES'S Explorations in S.W. Africa	15
BALL'S Guide to the Central Alps	15
———— Guide to the Western Alps	15
BAYLDON'S Rents and Tillages	12
BLACK'S Treatise on Brewing	19
BLACKLEY and FRIEDLANDER'S German and English Dictionary	5
BLAINE'S Rural Sports	17
BLIGHT'S Week at the Land's End	16
BONNEY'S Alps of Dauphiné	15
BOURNE'S Catechism of the Steam Engine	12
———— Handbook of Steam Engine	12
———— Treatise on the Steam Engine	11
BOWDLER'S Family SHAKSPEARE	17
BOYD'S Manual for Naval Cadets	18
BRAMLEY-MOORE'S Six Sisters of the Valleys	16
BRANDE'S Dictionary of Science, Literature, and Art	9
BRAY'S (C.) Education of the Feelings	7
———— Philosophy of Necessity	7
———— (Mrs.) British Empire	7
BREWER'S Atlas of History and Geography	19
BRINTON on Food and Digestion	19
BRISTOW'S Glossary of Mineralogy	8
BRODIE'S (Sir C. B.) Psychological Inquiries	7
———— Works	10
———— Autobiography	10
BROWNE'S Ice Caves of France and Switzerland	15
BROWNE'S Exposition 39 Articles	12
———— Pentateuch	12
BUCKLE'S History of Civilization	2
BULL'S Hints to Mothers	19
———— Maternal Management of Children	19
BUNSEN'S Analecta Ante-Nicæna	13
———— Ancient Egypt	2
———— Hippolytus and his Age	13
———— Philosophy of Universal History	13
BUNSEN on Apocrypha	13
BUNYAN'S Pilgrim's Progress, illustrated by BENNETT	11
BURKE'S Vicissitudes of Families	4
BURTON'S Christian Church	3
BUTLER'S Atlas of Ancient Geography	19
———— Modern Geography	19
Cabinet Lawyer	19
CALVERT'S Wife's Manual	14
Campaigner at Home	6
CATS and FARLIE'S Moral Emblems	11
Chorale Book for England	14
COLENSO (Bishop) on Pentateuch and Book of Joshua	13
COLUMBUS'S Voyages	16
Commonplace Philosopher in Town and Country	6
CONINGTON'S Handbook of Chemical Analysis	9
CONTANSEAU'S Pocket French and English Dictionary	5
———— Practical ditto	5
CONYBEARE and HOWSON'S Life and Epistles of St. Paul	12
COOK'S Voyages	16
COPLAND'S Dictionary of Practical Medicine	10
———— Abridgment of ditto	10
COX'S Tales of the Great Persian War	2
———— Tales from Greek Mythology	16
———— Tales of the Gods and Heroes	16
———— Tales of Thebes and Argos	16
CRESY'S Encyclopædia of Civil Engineering	11
Critical Essays of a Country Parson	6
CROWE'S History of France	2
D'AUBIGNÉ'S History of the Reformation in the time of CALVIN	2
Dead Shot (The), by MARKSMAN	17
DE LA RIVE'S Treatise on Electricity	8
DELMARD'S Village Life in Switzerland	15
DE LA PRYME'S Life of Christ	13
DE TOCQUEVILLE'S Democracy in America	2
Diaries of a Lady of Quality	3
DOBSON on the Ox	18
DOVE'S Law of Storms	7
DOYLE'S Chronicle of England	2

Edinburgh Review (The) 20
Ellice, a Tale........................... 16
ELLICOTT's Broad and Narrow Way........ 13
———— Commentary on Ephesians 13
———— Destiny of the Creature........ 13
———— Lectures on Life of Christ 13
———— Commentary on Galatians 13
———————————— Pastoral Epist. 13
———————————— Philippians,&c. 13
———————————— Thessalonians 13
Essays and Reviews 13
——— on Religion and Literature, edited by
MANNING 13
——— written in the Intervals of Business 6

FAIRBAIRN's Application of Cast and
Wrought Iron to Building............. 11
———————— Information for Engineers .. 11
———————— Treatise on Mills & Millwork 11
FFOULKES's Christendom's Divisions...... 13
First Friendship 16
FITZ ROY's Weather Book 7
FOWLER's Collieries and Colliers 12
Fraser's Magazine 20
FRESHFIELD's Alpine Byways............. 15
————————— Tour in the Grisons 15
Friends in Council 6
FROUDE's History of England............. 1

GARRATT's Marvels and Mysteries of Instinct 8
GEE's Sunday to Sunday 14
Geological Magazine 8, 20
GILBERT and CHURCHILL's Dolomite Mountains 15
GILLY's Shipwrecks of the Navy 16
GOETHE's Second Faust, by Anster........ 17
GOODEVE's Elements of Mechanism........ 11
GORLE's Questions on BROWNE's Exposition
of the 39 Articles 12
Graver Thoughts of a Country Parson 6
GRAY's Anatomy......................... 10
GREENE's Corals and Sea Jellies 8
————— Sponges and Animalculæ 8
GROVE on Correlation of Physical Forces .. 8
GWILT's Encyclopædia of Architecture 11

Handbook of Angling, by EPHEMERA...... 18
HARE on Election of Representatives 5
HARTWIG's Sea and its Living Wonders.... 8
——————— Tropical World 8
HAWKER's Instructions to Young Sportsmen 17
HEATON's Notes on Rifle Shooting 17
HELPS's Spanish Conquest in America 2
HERSCHEL's Essays from the Edinburgh and
Quarterly Reviews 9
——————— Outlines of Astronomy....... 7
HEWITT on the Diseases of Women 9
HINCHLIFF's South American Sketches.... 15
Hints on Etiquette 19
HODGSON's Time and Space............... 7
HOLLAND's Chapters on Mental Physiology 6
——————— Essays on Scientific Subjects.. 9
——————— Medical Notes and Reflections 10
HOLMES's System of Surgery............. 10
HOOKER and WALKER-ARNOTT's British
Flora 9
HORNE's Introduction to the Scriptures.... 13

HORNE's Compendium of the Scriptures .. 13
HOSKYNS's Talpa 12
How we Spent the Summer 15
HOWITT's Australian Discovery 15
——————— History of the Supernatural 6
——————— Rural Life of England 16
——————— Visits to Remarkable Places 16
HOWSON's Hulsean Lectures on St. Paul.... 12
HUGHES's (E.) Atlas of Physical, Political,
and Commercial Geography............ 19
——————— (W.) Geography of British History 7
——————— Manual of Geography 7
HULLAH's History of Modern Music 3
——————— Transition Musical Lectures 3
HUMPHREYS' Sentiments of Shakspeare.... 11
Hunting Grounds of the Old World 15
Hymns from Lyra Germanica............. 14

INGELOW's Poems 17
Instructor (The) 20

JAMESON's Legends of the Saints and Martyrs 11
——————— Legends of the Madonna 11
——————— Legends of the Monastic Orders 11
JAMESON and EASTLAKE's History of Our
Lord 11
JOHNS's Home Walks and Holiday Rambles 8
JOHNSON's Patentee's Manual 11
——————— Practical Draughtsman 11
JOHNSTON's Gazetteer, or Geographical Dictionary 7
JONES's Christianity and Common Sense .. 7

KALISCH's Commentary on the Old Testament 5
——————— Hebrew Grammar............. 5
KENNEDY's Hymnologia Christiana 14
KESTEVEN's Domestic Medicine 10
KIRBY and SPENCE's Entomology 9
KNIGHTON's Story of Elihu Jan 16
KÜBLER's Notes to Lyra Germanica...... 14
KUENEN on Pentateuch and Joshua....... 13

Lady's Tour round Monte Rosa 15
LANDON's (L. E. L.) Poetical Works....... 17
Late Laurels 16
LATHAM's English Dictionary 5
LECKY's History of Rationalism 2
Leisure Hours in Town 6
LEWES's Biographical History of Philosophy 2
LEWIS on the Astronomy of the Ancients .. 4
——— on the Credibility of Early Roman
History 4
——— Dialogue on Government........... 4
——— on Egyptological Method.......... 4
——— Essays on Administrations........ 4
——— Fables of BABRIUS............... 4
——— on Foreign Jurisdiction 4
——— on Irish Disturbances 4
——— on Observation and Reasoning in
Politics............................. 4
——— on Political Terms.............. 4
——— on the Romance Languages 4
LIDDELL and SCOTT's Greek-English Lexicon 5
——————— Abridged ditto 5
LINDLEY and MOORE's Treasury of Botany. 9

NEW WORKS PUBLISHED BY LONGMANS AND CO. 23

Longman's Lectures on the History of England 1
Loudon's Encyclopædia of Agriculture 12
——————— Cottage, Farm, and Villa Architecture 12
——————— Gardening 12
——————— Plants 9
——————— Trees and Shrubs 9
Lowndes's Engineer's Handbook 11
Lyra Domestica 14
—— Eucharistica............................ 14
—— Germanica 11, 14
—— Messianica 14
—— Mystica 14
—— Sacra 14

Macaulay's (Lord) Essays.................... 2
——————— History of England 1
——————— Lays of Ancient Rome........ 17
——————— Miscellaneous Writings 6
——————— Speeches 5
——————— Speeches on Parliamentary Reform............................. 5
Macdougall's Theory of War 12
Marshman's Life of Havelock 3
McLeod's Middle-Class Atlas of General Geography 19
——————— Physical Atlas of Great Britain and Ireland 19
McCulloch's Dictionary of Commerce.... 18
——————— Geographical Dictionary 7
Macfie's Vancouver Island 15
Maguire's Life of Father Mathew 3
——————— Rome and its Rulers 3
Maling's Indoor Gardener.................. 9
Massey's History of England.............. 1
Massingberd's History of the Reformation 3
Maunder's Biographical Treasury 4
——————— Geographical Treasury 7
——————— Historical Treasury 2
——————— Scientific and Literary Treasury 9
——————— Treasury of Knowledge........ 19
——————— Treasury of Natural History .. 9
Maury's Physical Geography.............. 7
May's Constitutional History of England .. 1
Melville's Digby Grand 16
——————— General Bounce 16
——————— Gladiators 16
——————— Good for Nothing 16
——————— Holmby House 16
——————— Interpreter 16
——————— Kate Coventry................. 16
——————— Queen's Maries 16
Mendelssohn's Letters 3
Menzies' Windsor Great Park 12
——————— on Sewage..................... 12
Merivale's (H.) Colonisation and Colonies 7
——————— Historical Studies 1
——————— (C.) Fall of the Roman Republic 2
——————— Romans under the Empire 2
——————— on Conversion of Roman Empire........................... 2
——————— on Horse's Foot................ 18
——————— on Horse Shoeing............. 18
——————— on Horses' Teeth 18
——————— on Stables 18
Mill on Liberty 4
—— on Representative Government 4
—— on Utilitarianism...................... 4
Mill's Dissertations and Discussions...... 4
——————— Political Economy 4
——————— System of Logic............... 4
——————— Hamilton's Philosophy........ 4
Miller's Elements of Chemistry........... 9
Monsell's Spiritual Songs 14
——————— Beatitudes................... 14
Montagu's Experiments in Church and State.................................. 13
Montgomery on the Signs and Symptoms of Pregnancy........................... 9
Moore's Irish Melodies 17
——————— Lalla Rookh 17
——————— Memoirs, Journal, and Correspondence 3
——————— Poetical Works............... 17
Morell's Elements of Psychology 6
——————— Mental Philosophy............. 6
Morning Clouds 14
Morton's Prince Consort's Farms 12
Mosheim's Ecclesiastical History........ 13
Müller's (Max) Lectures on the Science of Language 5
——————— (K. O.) Literature of Ancient Greece 2
Murchison on Continued Fevers.......... 10
Mure's Language and Literature of Greece 2

New Testament illustrated with Wood Engravings from the Old Masters 10
Newman's History of his Religious Opinions 3
Nightingale's Notes on Hospitals 19

Odling's Course of Practical Chemistry.... 9
——————— Manual of Chemistry 9
Ormsby's Rambles in Algeria and Tunis .. 15
Owen's Comparative Anatomy and Physiology of Vertebrate Animals 8
Oxenham on Atonement.................. 14

Packe's Guide to the Pyrenees 15
Paget's Lectures on Surgical Pathology .. 10
——————— Camp and Cantonment............ 15
Pereira's Elements of Materia Medica 10
——————— Manual of Materia Medica...... 10
Perkins's Tuscan Sculpture 11
Phillips's Guide to Geology 8
——————— Introduction to Mineralogy.... 8
Piesse's Art of Perfumery 12
——————— Chemical, Natural, and Physical Magic 12
——————— Laboratory of Chemical Wonders 12
Playtime with the Poets.................. 17
Practical Mechanic's Journal 11
Prescott's Scripture Difficulties 13
Proctor's Saturn 7
Pycroft's Course of English Reading 5
——————— Cricket Field 18
——————— Cricket Tutor 18
——————— Cricketana 18

Reade's Poetical Works 17
Recreations of a Country Parson, Second Series 6
Reilly's Map of Mont Blanc.............. 15
Riddle's Diamond Latin-English Dictionary 5
——————— First Sundays at Church 14
Rivers's Rose Amateur's Guide 9

Rogers's Correspondence of Greyson...... 6
——— Eclipse of Faith............ 6
——— Defence of ditto 6
——— Essays from the *Edinburgh Review* 6
——— Fulleriana 6
Roget's Thesaurus of English Words and Phrases 5
Ronalds's Fly-Fisher's Entomology 18
Rowton's Debater...................... 5
Russell on Government and Constitution. 7

Saxby's Study of Steam 18
——— Weather System 7
Scott's Handbook of Volumetrical Analysis 9
Scrope on Volcanoes 8
Senior's Biographical Sketches 4
——— Historical and Philosophical Essays........................... 3
——— Essays on Fiction............. 16
Sewell's Amy Herbert.................. 16
——— Ancient History................ 2
——— Cleve Hall 16
——— Earl's Daughter............... 16
——— Experience of Life 16
——— Gertrude 16
——— Glimpse of the World........ 16
——— History of the Early Church...... 3
——— Ivors 16
——— Katharine Ashton............. 16
——— Laneton Parsonage........... 16
——— Margaret Percival 16
——— Night Lessons from Scripture.... 14
——— Passing Thoughts on Religion... 14
——— Preparation for Communion... 14
——— Readings for Confirmation 14
——— Readings for Lent............ 14
——— Self-Examination before Confirmation............................. 14
——— Stories and Tales 16
——— Thoughts for the Holy Week.... 14
——— Ursula 16
Shaw's Work on Wine 19
Shedden's Elements of Logic 5
Short Whist 19
Short's Church History 3
Sieveking's (Amelia) Life, by Winkworth 3
Simpson's Handbook of Dining......... 18
Smith's (Southwood) Philosophy of Health 19
——— (J.) Voyage and Shipwreck of St. Paul................................. 13
——— (G.) Wesleyan Methodism 3
——— (Sydney) Memoir and Letters.... 4
——— ——— Miscellaneous Works .. 6
——— ——— Sketches of Moral Philosophy 6
——— ——— Wit and Wisdom 6
Smith on Cavalry Drill and Manœuvres.... 18
Southey's (Doctor) 6
——— Poetical Works............... 17
Spohr's Autobiography 3
Spring and Autumn 14
Stanley's History of British Birds........ 8
Stebbing's Analysis of Mill's Logic...... 5
Stephenson's (R.) Life by Jeaffreson and Pole 3
Stephen's Essays in Ecclesiastical Biography............................ 4

Stephen's Lectures on the History of France..................................... 2
Stepping Stone to Knowledge, &c.......... 20
Stirling's Secret of Hegel............. 6
Stonehenge on the Dog.................. 18
——— on the Greyhound 18

Tasso's Jerusalem, by James............ 17
Taylor's (Jeremy) Works, edited by Eden 14
Tennent's Ceylon...................... 8
——— Natural History of Ceylon 8
Thirlwall's History of Greece 2
Thomson's (Archbishop) Laws of Thought 4
——— (J.) Tables of Interest 19
——— Conspectus, by Birkett...... 10
Todd's Cyclopædia of Anatomy and Physiology 10
——— and Bowman's Anatomy and Physiology of Man 10
Trollope's Barchester Towers 16
——— Warden 16
Twiss's Law of Nations 18
Tyndall's Lectures on Heat.............. 8

Ure's Dictionary of Arts, Manufactures, and Mines 11

Van der Hoeven's Handbook of Zoology 8
Vaughan's (R.) Revolutions in English History 1
——— (R. A.) Hours with the Mystics 7
Villari's Savonarola 3

Watson's Principles and Practice of Physic 10
Watts's Dictionary of Chemistry......... 9
Webb's Celestial Objects for Common Telescopes 7
Webster & Wilkinson's Greek Testament 13
Weld's Last Winter in Rome............ 15
Wellington's Life, by Brialmont and Gleig 3
——— by Gleig 3
West on the Diseases of Infancy and Childhood 9
Whately's English Synonymes 4
——— Logic 4
——— Remains................... 4
——— Rhetoric 4
——— Sermons................... 14
——— Paley's Moral Phylosophy.... 14
Whewell's History of the Inductive Sciences.................................. 2
Whist, what to lead, by Cam 19
White and Riddle's Latin-English Dictionary.............................. 5
Wilberforce (W.) Recollections of, by Harford............................ 3
Williams's Superstitions of Witchcraft .. 6
Willich's Popular Tables 19
Wilson's Bryologia Britannica........... 9
Wood's Homes without Hands........... 8
Woodward's Historical and Chronological Encyclopædia 2

Yonge's English-Greek Lexicon 5
——— Abridged ditto 5
Young's Nautical Dictionary............. 18
Youatt on the Dog 18
——— on the Horse 18

www.ingramcontent.com/pod-product-compliance
Lightning Source LLC
Chambersburg PA
CBHW032011220426
43664CB00006B/214